Run Your Diesel Vehicle on Biofuels

Run Your Diesel Vehicle on Biofuels

A Do-It-Yourself Guide

Jon Starbuck and Gavin D. J. Harper

McGraw Hill

New York   Chicago   San Francisco   Lisbon   London   Madrid
Mexico City   Milan   New Delhi   San Juan   Seoul
Singapore   Sydney   Toronto

The McGraw·Hill Companies

Library of Congress Cataloging-in-Publication Data

Starbuck, Jon.
     Run your diesel vehicle on biofuels : a do-it-yourself manual / Jon Starbuck
and Gavin D. J. Harper.
          p.   cm.
     Includes bibliographical references and index.
     ISBN 978-0-07-160043-9 (alk. paper)
          1. Motor vehicles—Motors (Diesel)—Amateurs' manuals. 2. Motor vehicles—
Motors—Modification—Amateurs' manuals. 3. Diesel automobiles—Customizing—
Amateurs' manuals. 4. Diesel motor—Alternative fuels—Amateurs' manuals.
5. Biodiesel fuels—Amateurs' manuals. 6. Do-it-yourself work—Amateurs' manuals.
I. Harper, Gavin D. J. II. Title.
     TL229.D5S73   2009
     629.25'06—dc22   2008044955

1 2 3 4 5 6 7 8 9 0 DOC/DOC 0 1 4 3 2 1 0 9 8

ISBN 978-0-07-160043-9
MHID 0-07-160043-4

Sponsoring Editor: Judy Bass
Production Supervisor: Richard C. Ruzycka
Editing Supervisor: Stephen M. Smith
Project Manager: Andy Baxter, Keyword Group Ltd.
Copy Editor: David Burin
Proofreader: Barbara Danziger
Indexer: Golden Paradox Indexing
Art Director, Cover: Jeff Weeks
Composition: Keyword Group Ltd.

Printed and bound by RR Donnelley.

McGraw-Hill books are available at special quantity discounts to use as premiums
and sales promotions, or for use in corporate training programs. To contact a special
sales representative, please visit the Contact Us page at www.mhprofessional.com.

This book is printed on acid-free paper.

About the Authors

Jon Starbuck makes his own biodiesel and converts vehicles to run on his cleaned-up chip-shop oil. He writes websites and teaches courses on the subjects of renewable energy, vegetable oil, and biodiesel. He holds a B.Sc. in Physics from the University of Kent at Canterbury (UK) and is currently pursuing an M.Sc. in Renewable Energy Systems Technology at Loughborough University (UK). He lived, worked, and taught at the Centre for Alternative Technology (CAT) in Wales for five years, and he is now a freelance renewable energy engineer. He still lives in Wales with his wife and their dog, where he is always making something. This is his first book.

www.jonstarbuck.co.uk

Gavin D. J. Harper is the author of *50 Awesome Auto Projects for the Evil Genius, Build Your Own Car PC, 50 Model Rocket Projects for the Evil Genius, Solar Energy Projects for the Evil Genius,* and *Fuel Cell Projects for the Evil Genius* (all from McGraw-Hill), and has had work mentioned in the journal *Science.*

He holds a Diploma in Design and Innovation and a B.Sc. in Technology from the Open University, UK. He went on to study toward an M.Sc. in Architecture with Advanced Environmental and Energy Studies with the University of East London at the Centre for Alternative Technology, Wales. He also holds the Diploma of Vilnius University, Lithuania. He has undertaken further study with the Open University and with Loughborough University's Centre for Renewable Energy Systems Technology. He is currently reading for his Ph.D. into the impacts of alternative vehicles and fuels at Cardiff University, Wales.

Contents

Contents

Foreword

I can remember the first time I met someone with a vegetable oil ride. A classmate's boyfriend had done a full "waste vegetable oil" (WVO) conversion of his decade-old F-150. I honestly couldn't believe what I was hearing. I must have grilled him about it for half an hour. How dirty is the exhaust? Can you start it in cold weather? How do you have to prep the oil? How long did it take you to get it up and running?

And finally, after he'd answered all those questions for me, and I was satisfied that waste vegetable oil was both a viable and environmentally friendly option, I asked him whether he could make one for me. Unfortunately, this was the one question I asked that had an unsatisfying answer. He didn't have the time to make WVO cars for other folks. And so, to this day, I'm driving a regular, old boring gasoline car.

But the possibilities of WVO-powered travel are considerably more mouth-watering than the fried food responsible for the fuel itself. I may want a chicken nugget every once in a while, but I want free gasoline every single day of my life. And don't make a mistake, that's what we're talking about here … free, clean fuel, straight from the trash. And while I never had the expertise (or a willing friend) to build a biodiesel vehicle for myself, now Jon and Gavin have provided a guide for us all.

But don't mistake, you're not going to be one of billions traveling on this fuel. There simply isn't enough waste oil out there to power much of our current fleet. The path you're about to go down isn't going to solve the world's energy crisis, but it is an important step. Simply noticing that this valuable fuel has been treated like toxic waste for the last 50 years is a great step on the path to a more efficient world. It's people like Gavin, Jon, and (hopefully) you who can identify the profligate waste of our systems and use it to their own advantage, and who will teach the rest of humanity how to behave on a finite planet.

HANK GREEN
Editor-in-Chief/Founder
EcoGeek.org

Hank Green is the Editor-in-Chief and Founder of EcoGeek.org, the largest environmental technology publication on the Internet. If you want to learn about the thousands of ways in which the world's smartest people have committed themselves to growing our civilization without destroying the planet, you should check it out.

Preface

Before you begin

Before you so much as read the Introduction to this book, we want to make sure that we are all singing from the same song-sheet. Some of the terms we are going to use in this book are often misunderstood. Journalists who don't understand the technology writing articles from corporate press releases, a lack of "green literacy" in the media reporting stories on sustainability, and just plain confusion means that terms are often used loosely and indiscriminately. We've tried to be as consistent as possible in this book to avoid confusion, so read the following before you start so that we are all using the same terms of reference!

What is in a name? Clarifying terminology and biofuels basics

There are a number of reasons why the biodiesel terminology has become muddled – from a simple misunderstanding to the very deliberate campaigns by some in the biofuels community to confuse the public by using the same words to mean different things. Here we are going to explain what *we* mean by various terms in this book *and* terms that you may come across in other publications and across the Internet.

Be vigilant and mindful at all times when reading information on biofuels and remember:

> There are some that only employ words for the purpose of disguising their thoughts.
>
> Voltaire (1694–1778), Dialogue XIV, *Le Chapon et la Poularde* (1766)

Biodiesel is a fuel *made from* vegetable oil and/or animal fat. It is not vegetable oil, and has been chemically altered to make it more like mineral-diesel fuel. Most commonly, the vegetable oil is reacted in the presence of a catalyst with methanol to make fatty acid methyl ester, or FAME.

Biodiesel blends are not 100% biodiesel. Biodiesel refers to the pure fuel. Biodiesel blends, "green" diesels, and so on are usually denoted as "BXX," where "XX" represents the percentage of biodiesel contained in the blend; B20 is 20% biodiesel and 80% mineral diesel, and so on.

SVO or straight vegetable oil is just that: vegetable oil, usually food-grade cooking oil, unused and unmodified, such as you can buy in any supermarket or from any wholesaler. Some in the industry call it PPO, pure plant oil.

WVO or UCO or grease is waste cooking oil, used cooking oil, yellow grease. It is the vegetable oil and/or animal fats that have been used by a restaurant and are now a waste product. It is usually dirty, has some water in it, and is not very pleasant to handle. This is the typical raw ingredient for making biodiesel.

Diesel or mineral-diesel petrodiesel or even dinodiesel is the traditional diesel fuel you would expect to find on a filling station forecourt; it is also sometimes referred to as DERV. It is a distilled fraction of crude oil that was formed over millions of years in the earth from dead plant matter, usually algae (sadly not dinosaurs). Modern mineral-diesel has a large number of additives in it.

MWVO is modified waste vegetable oil. Confusingly, it can mean a number of things, but usually refers to waste vegetable oil that has had a thinning agent added to it to reduce its viscosity rather than being chemically modified.

RUG or regular unleaded gasoline is the fuel sold on gas station forecourts across the world for use in modern gasoline engines. It is unleaded gas (USA) or unleaded petrol (UK).

Other terms used in the book (see also Appendix) are **RME**, rapeseed methyl ester, a specific form of biodiesel made from rape or canola oil; **PPO**, pure plant oil (see SVO or straight vegetable oil); **ULSD**, ultra low sulphur diesel; **CO**, carbon monoxide; **NO$_x$**, various oxides of nitrogen; **CNG**, compressed natural gas; **FAME**, fatty acid methyl ester, biodiesel.

Acknowledgments

The authors would like to thank the following people who have contributed to the development of this book.

A massive thank you to the two Peters: Hampson and Jackson, of City Yacht School, UK. Peter Hampson was especially invaluable in sharing his knowledge about diesel engine technology, and kindly let us include pictures of some of his diesel engine teaching aids, and an eclectic collection of diesel engine parts and tools, in this book. Geoff Harper helped to source diesel pictures and put us in touch with the right people. Hearty thanks to Charles J. Melton and Steven Charles at Cummins Inc. for their help in sourcing diesel engine cutaways for the chapter on diesel engine technology.

Gavin would like to thank Willie Nelson for giving him a platform in his fabby little book *On the Clean Road Again* – a must read. A massive thanks to Annie Nelson, Bob and Kelly King, and all the folks at Pacific Biodiesel for their warm welcome to the 2007 National Biodiesel Conference – these folks champion the Sustainable Biodiesel Alliance (SBA; www.sustainablebiodieselalliance.com), which aims to keep a bit of integrity in the biodiesel industry by keeping it green, clean, local, and free from GM, nasty additives, colors, sweeteners, and artificial flavors. Without the invite to the SBA, Gavin would never have met Frankie Abralind, editor of the truly awesome *BiodieselSMARTER*, a smashing little quarterly, for which you can find a discount coupon in the back of this book.

An especially big thank you to Amanda Starbuck, Jon's wife, for the financial and emotional support and the beer. Thanks to Robert Starbuck, Jon's brother, for his photos and prompt responses to emails. Big thanks to Greenpeace, especially Timo for the desk space and the coffee, and special thanks to Gesche for keeping Jon hydrated. Thanks to the city of San Francisco for hosting Jon while he typed the manuscript and cycled about like a maniac in a futile attempt to work off the burritos.

A few people in the UK biofuels community have been especially helpful: big thanks to Jon and Dan at Golden Fuels (Oxford, UK) as well as Jan at Sundance Renewables (Swansea, UK); thanks also to Patrick Whetman of Goat Industries, and to Daniel Blackburn of www.vegoilmotoring.com for the UCO and for dressing up as Elvis.

The great people at the Centre for Alternative Technology (CAT), Wales, introduced the authors; thank you to the folks of the CAT kitchen, especially Yvonne, Natalia, and Dave, for the smashing chickpea curries that will always hold a special place in Gavin's heart.

Thanks to the Solar Living Institute in Hopland, California, especially Ashley Schaeffer and Pete Huff for putting Jon and Amanda up and showing them about, and John Schaeffer for the delicious breakfast.

Thanks also to the following people: Andrew Morris in Oakland, California, for showing Jon his set-up; Maria Alovert, aka Girl Mark, for her book on biodiesel that is probably the best homebrew book available; and Graham Laming, for brilliant ideas and a ton of free, high-quality information on the web (if you find any of Graham's information useful please make a donation to Cancer Research UK; details in the back of this book).

Our thanks to the people of Machynlleth and the Dyfi Valley who are dear to both the authors, and to Ben Robinson (who's filling up a truck in Chapter 12) and all at Dulas for letting Gavin take pictures of their big plastic tank.

Thanks to Roustabout Ltd, especially Geoff. Thanks for the letters and see you at Burning Man.

Thanks to Spanner Films and *The Age of Stupid* – go see it.

Remember, open source not secret source!

This book was brought to you by the letter B and the number 100.

Run Your Diesel Vehicle on Biofuels

Chapter 1

Introduction

Finding ways to run your diesel engine on fuels derived from vegetable, or indeed animal, oils is an exciting hobby, and one which can help reduce the cost of running a vehicle, and if done correctly, improve your carbon footprint.

There is a wealth of different literature, in print and on the Internet, covering a range of different solutions and a plethora of different approaches and techniques: In this book we hope to unify the existing literature on the topic, bringing together the different solutions, and helping you, the reader, to decide which one, if any, is for you.

Throughout this book, we have tried to maintain a tone of being critical about the technology, rather than being evangelical. We do not maintain that biofuels are right for every application, nor are some of the techniques used to produce it "sustainable." We encourage you to become widely read on the subject, using this book as a stepping stone to finding out much more about the topic. Throughout the book, we try to flag up useful web links and further sources of information. We encourage you to investigate the sources of further reading and information, and to make your own informed conclusions.

The information in this book represents our attempt to bring together the current "practical" literature relating to biofuels. This is not an academic book, although academic sources are cited. It does not attempt to be a comprehensive evaluation of the academic literature that is developing on the subject; it is, however, a thorough, practical handbook.

Biofuels are such polarized subjects at the moment, with good arguments on both sides of the fence. We believe that a full understanding of biodiesel is only possible with an understanding of where the technology has come from, the social and ethical implications of using the technology, and a full understanding of a range of techniques and applications.

We live in a world that is facing a whole catalog of environmental problems. Additionally, with the prospect of resource scarcity, the global economy will be forced (and indeed is at the moment) adapting to change. At this stage, where we are looking for new technologies and solutions, it is too early to write off any technology that shows hope and promise; however, as we investigate the technology, we must do so slowly and cautiously and not always look to technology for solutions that can be reached more easily by social adaptation and changing lifestyles.

The birth of biodiesel

On August 31, 1937, G. Chavanne, a Belgian academic at the University of Brussels, obtained a patent, entitled "Procedure for the transformation of vegetable oils for their uses as fuels" – Belgian Patent 422,877. This patent described the alcoholysis – which in Chapter 5, we refer to as transesterification – of vegetable oils using ethanol (and mentions methanol) in order to separate the fatty acids from the glycerol by replacing the glycerol with short linear alcohols. This appears to be the first account of the production of what is known today as "biodiesel."

We are going to hold up on giving you a full account of the history and chemistry of biodiesel for now, which is covered in later chapters; for the time being, we are going to try and capture your imagination for the subject by explaining why there is a need for this technology.

Vehicle emissions and biodiesel

Vehicles produce large amounts of localized pollution. You only have to look at pictures of cities like Los Angeles and Mexico City to realize what a tremendous impact smog has on local air quality. Compared to petrodiesel, biodiesel offers a cleaner solution at the point of combustion and can readily be used in present-generation vehicles.

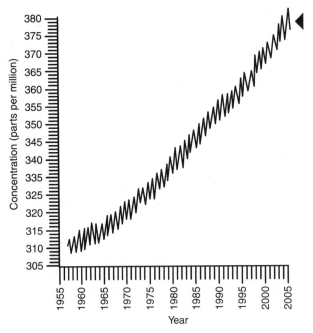

Figure 1.1
Atmospheric carbon dioxide levels measured at Mauna Loa Observatory.

To understand vehicle emissions, let us take a look at how vehicle pollution is formed.

The high temperatures that are reached inside a diesel engine during the process of combustion cause the nitrogen in the air to react with oxygen in the air to produce nitrogen oxide as an exhaust pollutant. This process is called "photolysis":

$$N_2(g) + O_2(g) \rightarrow 2NO(g)$$

Nitrogen oxide can in turn react with oxygen to produce nitrogen dioxide. We often refer to a mixture of nitrogen oxide and nitrogen dioxides as "NO_x," which stands for nitrogen dioxides. This is only an abbreviation; it is not a proper chemical symbol.

$$2NO(g) + O_2(g)2 \rightarrow NO_2(g)$$

Sunlight often causes one of the oxygen atoms to split away from the nitrogen dioxide molecule (remember, this is a "single" oxygen atom, not "diatomic" oxygen, the form in which it exists as a gas); however, this isn't the end of the story:

$$NO_2(g) \xrightarrow{\text{sunlight}} NO(g) + O(g)$$

This single atom of oxygen combines with diatomic oxygen in the air to form another variant of oxygen, namely ozone:

$$O + O_2 \rightarrow O_3$$

Ozone is an irritant. No, not the Moldovan pop music trio, the gas. As we have seen from the chemistry above, vehicle emissions lead to the production of ozone. If you've ever noticed a strange smell when standing on a subway platform, or after a lightning storm have smelled that peculiar odor in the air, the chances are you've smelt ozone. In fact Schonbein, who discovered ozone, named it after the Greek word *ozein* for odor.

In addition, single oxygen atoms liberated from ozone can then combine with nitric oxide (NO) to form nitrogen dioxide and diatomic oxygen:

$$NO(g) + O_3(g) \rightarrow NO_2(g) + O_2(g)$$

We get more nasty pollutants, when nitrogen dioxide, oxygen, and unburnt hydrocarbons from the exhaust (because the engine can't hope to burn 100% of the fuel) all combine together in sunlight:

$$NO_2(g) + O_2(g) + \text{hydrocarbons} \xrightarrow{\text{sunlight}} CH_3CO\text{-}OO\text{-}NO_2(g)$$

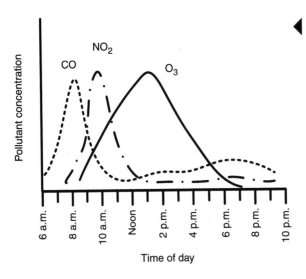

Figure 1.2
Evolution of pollutants during a smoggy day in Los Angeles.

They form peroxyacyl radicals, which combine with nitrogen to make peroxyacetyl nitrates (PANs), which cause irritation to the eyes and the respiratory tract. One of the reasons that they are so irritating to the human body is because they dissolve readily in water. The eyes and respiratory tract both provide sources of moisture in which the PANs will readily dissolve, causing irritation and discomfort.

Our ignorance is not so vast as our failure to use what we know.
M. King Hubbert

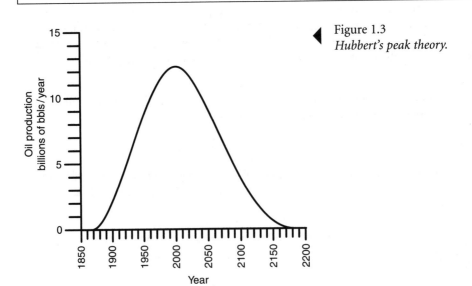

Figure 1.3
Hubbert's peak theory.

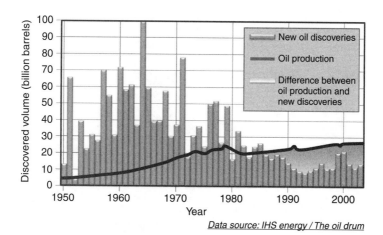

Figure 1.4
Oil discoveries and oil consumption compared.

If we look at a gallon of diesel, we can begin to see where the money goes. 2005 figures show that the average cost of a gallon of diesel was around $2.40.

The oil companies are giant monoliths, with many employees to support.

If we take the United States as an example, we can see that 21% of the cost of a gallon of diesel will go to taxes. Some European countries have much higher taxing regimes; as a result, a larger proportion of the cost of a gallon of diesel in these countries goes to pay for taxes.

Eight percent of the cost of every gallon of diesel that goes in your tank goes to marketing and distribution. When you think about the massive operation needed to ship diesel all over the world and to advertise and sell it, you begin to realize that with a small-scale, community or home biodiesel production plant, this cost can be massively reduced.

Crude oil is turned into a variety of petrochemical products, by a process known as fractional distillation. Crude oil contains a number of hydrocarbons, which are of different lengths. The different fractions all have different boiling points. In fractional distillation, the crude oil is heated, and the vapor allowed to rise through a column, which, uncannily, is called a fractional distillation column or fractionation tower.

Hydrocarbon boiling points follow a relatively simple set of rules. We are going to learn a lot more about hydrocarbons later in this book in Chapter 5, but all you need to know for the time being is that fractions with the higher boiling points:

• Have more carbon atoms in the chain
• Are heavier (having a higher molecular weight)
• Are thicker (more viscous)

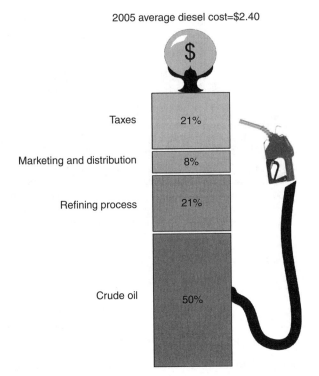

Figure 1.5
How the cost of diesel at the pump breaks down.
Source: Department of Energy – Energy Information Administration – Diesel Fuel Prices:
www.eia.doe.gov/bookshelf/brochures/diesel/dieselprices2006.html.

- Are harder to ignite
- Are darker in color

Contrasting gasoline (petrol) with diesel for a moment, we can see how diesel is heavier and thicker than gasoline, harder to burn (which is why it won't work in a spark-ignition engine), and is a darker color than gasoline. This is a property of the longer hydrocarbon chain length, which is one of the distinguishing characteristics of diesel over gasoline.

Back to our fractional distillation columns, which have a number of internal baffles, and form a series of zones of different temperature ranges within each column. Heat is applied at the bottom of the column, so that the bottom of the column is very hot while the top of the column is relatively cool.

The lighter fractions pass through the series of baffles to the top of the column and, where the temperature is cool, they condense. Meanwhile, the heavier fractions are "tapped off" at lower points on the column.

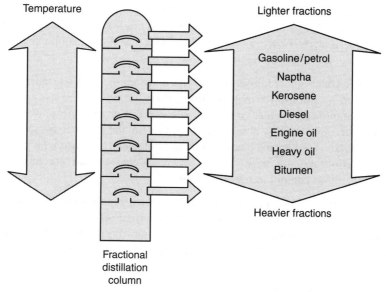

▲ Figure 1.6
Fractional distillation column.

Fractional distillation is generally a continuous process, with new fresh crude oil being input into the process and derivative products being steadily drawn off from the column.

Diesel fuel makes up one of the fractions of crude oil; for every 42-gallon barrel of crude oil, we can expect to extract 7.8 gallons of diesel, as shown in Figure 1.7.

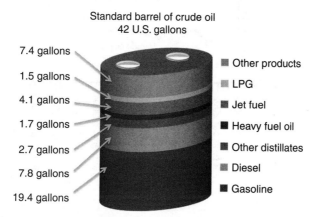

▲ Figure 1.7
How a barrel of crude breaks down into its fractions.
Source: Department of Energy – Energy Information Administration – Diesel Fuel Prices:
www.eia.doe.gov/bookshelf/brochures/diesel/dieselprices2006.html.

This results in the price of diesel tracking the price of a gallon of crude oil. The two are tied by the ability to supply this commodity: If there is less crude oil available, there will be less diesel produced. However, basic economics tells us that it is not just supply that governs the price of a commodity – it is also demand.

Demand for diesel has been steadily growing. With more advanced diesel engine technologies being introduced by automotive manufacturers, there are now many less concerns about the performance of diesel engines compared with gasoline engines. Time was, when diesel engines were the noisy, smelly, clunky poor relation in the internal combustion engine family; however, advanced engine management technologies, combined with more sophisticated fuel injection systems, have resulted in diesel engines that are relatively sedate, compared to the diesel engines in times gone by.

Luxury cars in continental Europe have employed diesel engines for some years as, one by one, car manufacturers awaken to the sophistication of modern diesel engines. Even the British car manufacturer Jaguar, a marque that has become synonymous with refined performance and understated class, now offers a diesel engine variant of its luxury saloons – long-held

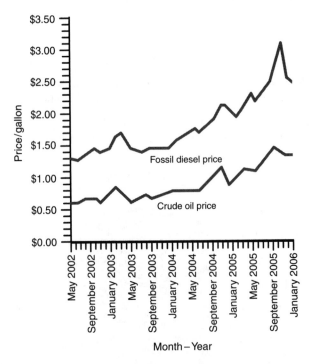

Figure 1.8
How the price of fossil diesel tracks crude oil prices.
Source: Department of Energy – Energy Information Administration – Diesel Fuel Prices:
www.eia.doe.gov/bookshelf/brochures/diesel/dieselprices2006.html.

perceptions about the diesel engine have been shattered with the advancement of technology, leading to a much greater demand for diesel fuel. In an age where oil is expensive, car buyers prize the higher efficiency of diesel engines, resulting in a greater number of miles per gallon – far easier on the wallet!

Furthermore, the taxation regimes in different countries can sometimes favor diesel fuel – while taxes are transient and often changing, where there has been a surplus of diesel fuel in the past, this is often reflected in a cheap taxation regime to encourage its use. However, as more and more drivers switch to diesel, the need for this tax incentive is diminished.

Introducing ... biofuels

So, with present-generation internal combustion engine technology, how can we escape the need for fossil fuels? The question is not an easy one to answer – someone more evangelical than we might shout out "biofuels," but as we hope the later chapters in this book will reinforce, biofuels are only part of the solution. However, if we bear in mind sensible considerations about biofuel production, and pursue the technology with a mind to reducing waste and minimizing our environmental impact rather than carrying on with the status quo, then in some circumstances, biofuels can be an appropriate energy answer.

The premise on which the environmental benignity of biofuels is based is this: Green plants operate using a biochemical process called photosynthesis (Figure 1.9). As well as obtaining

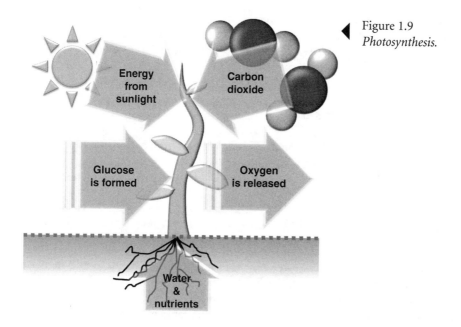

Figure 1.9
Photosynthesis.

Energy from sunlight

Carbon dioxide

Glucose is formed

Oxygen is released

Water & nutrients

nutrients from soil (or agricultural chemicals), plants take in water from the ground, with energy from sunlight, and carbon dioxide from the air, they synthesize the building blocks for plant cell structures and food. As plants "inhale" carbon dioxide, they "exhale" oxygen – the opposite of us mammals. Therefore, the carbon dioxide that the human race and animal kingdom produce naturally, is in part offset by plants and other flora taking this carbon dioxide and turning it back into oxygen – our plants act as the "lungs" of the world.

Now, the carbon from the atmosphere is locked in the cellular structure of the plants that grow with energy from the sun. However, when these plants are ready for harvest, we can process them and extract oil from them, by milling them, crushing them, and sieving what is left (we discuss the chemistry of oils in Chapter 5). What we are left with is "trapped carbon" from the atmosphere, which is passed through the process of growing the plant, harvesting it, and turning it into oil. We can then take the oil, and burn it directly, as you will see in later chapters, or chemically convert it into an analogue of mineral diesel, albeit one that has been "biologically produced." Brighter readers by this point should have realized that this is why it is called biodiesel.

However, when we burn this fuel, we release the carbon that was temporarily trapped in the plants and later the oil, back into the atmosphere – for plants to then reabsorb as they grow. However, as we will see, this is not the end of the story.

The great industrialist Henry Ford believed in the potential of biofuels. When he designed the Model-T Ford motor car, the car that brought automobility to the masses, he designed it so that it had the potential to run on bioethanol. Ford himself is quoted as saying:

the fuel of the future is going to come from apples, weeds, sawdust – almost anything. There is fuel in every bit of vegetable matter that can be fermented, enough alcohol in one year's yield of an acre of potatoes to drive the machinery necessary to cultivate the fields for a hundred years.

Henry Ford (1863 – 1947)

Emissions from biodiesel

The emissions from biodiesel are better in many respects than from petrodiesel. Testing confirms that, when burned in similar conditions, biodiesel results in fewer unburned hydrocarbons, less carbon monoxide, and a decreased count of particulate matter. This is evidenced in Figure 1.10.

Walker, W. "Biodiesel from Rapeseed." *Journal of the Royal Agricultural Society of England.* 1994. 155, pp. 43–44.

A desire for energy independence, a search for more sustainable solutions, not to mention the amount of money that can be made, has led to a rapid increase in global biodiesel production in the past decade. The U.S. case indicates this, with Figure 1.12 drawn from statistics from the U.S. National Biodiesel Board that illustrates the rapid growth of this technology.

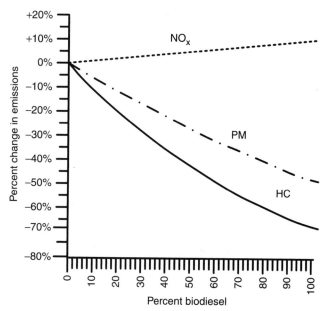

Figure 1.10
Emissions from biodiesel. PM, particulate matter; NO$_x$, nitrogen dioxide; HC, hydrocarbons.

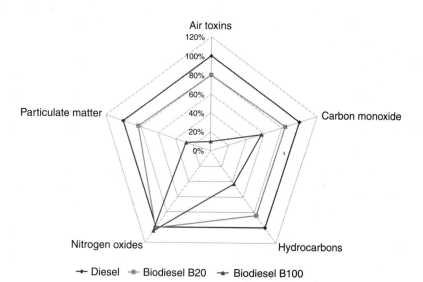

Figure 1.11
Radar chart showing comparison of emissions from diesel, B20 biodiesel blend and B100 biodiesel.

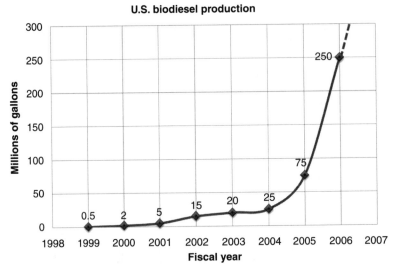

Figure 1.12
U.S. biodiesel production.
Source: Estimates from the U.S. National Biodiesel Board.

However, not everyone in the biodiesel industry is producing a product that can be classed as "sustainable." In Chapter 14 on Biofuel Ethics, we discuss some of the dilemmas surrounding biodiesel production.

Biomass, in which we class energy derived from burning plants, has always been an important energy source to mankind. Before the industrial revolution and widespread use of coal as a fuel, wood was the most commonly used fuel: bits of log gathered from the woods, or heated in the absence of oxygen to produce charcoal. Biomass still plays an important part in man's energy portfolio, as evidenced in Figure 1.13, from United Nations statistics.

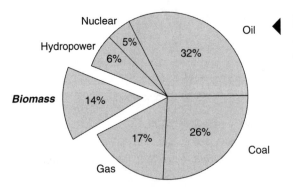

Figure 1.13
Hierarchy of global energy sources:
Source data: GREEN ENERGY UN.

The carbon dioxide produced by burning fossil fuel diesel slowly accumulates in our atmosphere – slowly choking the planet

Figure 1.14
Emissions from burning conventional petrodiesel.

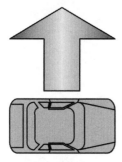

We burn fossil fuel diesel in our cars and vehicles. This produces a whole host of emissions

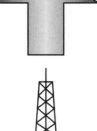

Oil is extracted from the ground and refined into mineral diesel. Once the fossil fuels are exhausted, they will take millions of years to form again

In fact, we can see that biomass and large-scale hydropower, while contributing significantly to world energy sources, are dwarfed by the massive amounts of fossil fuel and unsustainable nuclear power that we consume, all resulting in waste products that our environment must deal with.

When burning conventional fossil fuels, carbon is passed through a linear chain of being extracted from the ground, processed, and burned in the internal combustion engines of our vehicles, before being released into the air. It is an open cycle, a pattern of consumption, with no renewal involved. Consequently, resources are extracted, burned, and the products end up in our atmosphere, where they stay. Fossil fuels take millions of years to form from plant

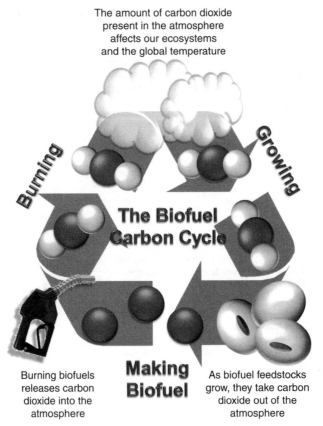

The amount of carbon dioxide
present in the atmosphere
affects our ecosystems
and the global temperature

Burning

Growing

The Biofuel
Carbon Cycle

Making
Biofuel

Burning biofuels
releases carbon
dioxide into the
atmosphere

As biofuel feedstocks
grow, they take carbon
dioxide out of the
atmosphere

Figure 1.15
The "idealized" biofuel carbon cycle.

matter compressed by layers of rock; unfortunately, the rate at which we release the carbon into the atmosphere is not matched by the rate at which new carboniferous matter is sequestered by dead plant matter and turned into fossil fuels – so there is no "cycle"; the process is linear.

Proponents of biofuel would like you to believe that emissions from biofuel are part of a "closed carbon cycle," a message reinforced in the National Biodiesel Board's FAQs; however, this is, at best, an oversimplification of the truth. In an idealized biofuel closed cycle, plants capture carbon dioxide from the atmosphere, as plants take in carbon dioxide, water, sunlight, and nutrients from the soil, using a process called photosynthesis. However, in reality, ideal models rarely represent the full complexity of the processes occurring in the real world. To view the carbon input into biofuel production as a closed cycle is at best a distortion of

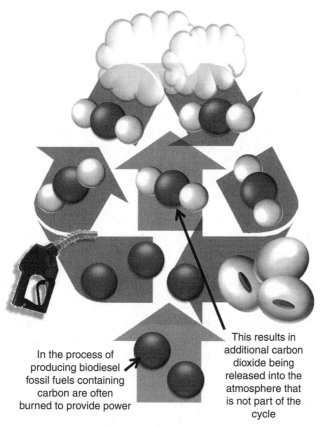

In the process of producing biodiesel fossil fuels containing carbon are often burned to provide power

This results in additional carbon dioxide being released into the atmosphere that is not part of the cycle

Figure 1.16
The biofuel carbon cycle in reality.

the truth, and at worst a real stinking lie! We're going to explore some of the other inputs needed for a sustainable biofuel production cycle in Chapter 14, and look at some of the arguments surrounding the sustainability of biofuel production, but for the time being, let's conduct a thought experiment.

Let's think about growing things commercially. It's not just a case of planting the seeds and watching them grow: Energy is needed to produce seedstock on a commercial scale and transport it to the farm. Fuel is then consumed by the agricultural machinery that sows the seeds; there is also energy input required to plough the ground to prepare for sowing. This is not insignificant. Think about the effort required to dig a hole in your yard: You quickly work up a sweat!

Then, once the seed is planted, there is energy required to pump water for irrigation. In commercial agriculture, pesticides and other agrichemicals are also needed; these are produced from chemical feedstocks that need to be mined from the earth and processed into a form that is suitable for use in agriculture. Producing these chemicals requires inputs of energy at all stages of the supply chain.

Then there is also the case of harvesting the crops once they have grown, which again requires agricultural machinery. The processing of crops, by milling or crushing, also requires an input of energy. Subsequently, when the oil has been extracted, we still need to heat it, add methoxide (see Chapter 5), and perform processes on the resulting mixture to obtain those esters which eventually go into your vehicle. This all requires energy, so we need to be careful that the sums add up – i.e., the amount of energy we expend making the stuff is lower than the energy we get out.

Of course, there are plenty of opportunities to substitute renewable sources of energy for fossil fuel sources of energy at all stages of this supply chain: For example, delivery vehicles running on biofuel or electric machinery running on renewable energy would help us to reduce the carbon intensity of the process. However, we can only make these changes to the process if we are mindful of what is happening and take affirmative action to improve the sustainability of our processes!

It has been said that "commercial agriculture is just a system for turning oil into food." If the product of a portion of our agricultural business is biofuel not food, we have to question the effectiveness of turning fossil fuel into biofuel, with efficiency losses in the process. The only way we can make the system sustainable is if in the process we can capture *more energy* (from the sun through photosynthesis) and *more carbon* (through plants growing and capturing carbon dioxide) than we put into the process from other inputs. If we can get our feedstock from a waste resource, such as used oil which would be sent to landfill, then we can make a net energy gain.

The lowdown on biodiesel maladies

Rudolf Diesel intended his engine to be adaptable and to run on a variety of fuels; however, just about all modern diesel engines are designed to run on modern, mineral diesel fuel oil and are not very tolerant of being run on vegetable oil. This can be summed up in one word – viscosity.

Vegetable oil, whether it is new or used, is significantly more viscous, thicker or less runny, than mineral diesel. One way or another we have to get our vegetable oil to be a similar viscosity to the mineral diesel fuel the engine was designed for; there are several different approaches.

It is usually at this point that someone says that they know this guy who has been putting neat vegetable oil into his car for years and has never had any problems. Indeed, one of the authors is guilty of putting no end of different oils in his tank and has always appeared to get away with it. Just as we all know someone who knows someone who smoked 40-a-day and lived until they were 100-years-old and yet we still know smoking is bad for you, putting unprocessed oil into an unmodified vehicle will lead to problems sooner or later.

Indeed, this may in part be a product of the confusion in the public's mind over the difference between biodiesel and vegetable oil and in biodiesel damaging engines. In Greg Pahl's book *Biodiesel*, Raffaello Garofalo of the European Biodiesel Board points out that "one barrel of bad Biodiesel is enough to spoil the reputation of the entire biodiesel industry." He goes on to write about people who put vegetable oil into their cars, which, after a while, caused the engine to fail, the cars being returned under warranty to the manufacturers with the claim that the biodiesel had caused the failure!

We're going to show you how to do things properly in this book. We focus on tried and tested methods for home biodiesel production and methods for converting your vehicle to run on SVO and WVO. If this sounds like too much hassle by the time you've finished this book, then there are plenty of places where you can now obtain biodiesel or other diesel fuels "off the shelf."

Great ... but I'm skeptical ... nowhere I know sells biodiesel

It's one thing finding a great local supplier of biodiesel who you know produces good-quality fuel, but what if you want to make a long journey on biofuel and you're concerned about finding refueling locations. Got an iPhone, Blackberry, or a PDA? Then use Earthcomber to locate your local friendly biofuel outlet! It's a free PDA application that will help you to locate alternative fuel sources. Earthcomber contains not only locations for biodiesel but also for the following fuel types:

- Biodiesel
- E85 (ethanol)
- LPG (liquefied petroleum gas)
- CNG (compressed natural gas)
- Hydrogen
- Electric (hookups to recharge electric cars and hybrids)

Earthcomber software: www.earthcomber.com.

Final remarks

We hope that you enjoy the chapters that follow. We've tried to be as comprehensive as possible, and cover the subject from all angles, not just the technology; however, we also give you an overview of the politics and the social scenarios that surround biofuels. We urge you to read more about the subject; we've tried to provide as many leads and references as we can for you to hunt out further information. We hope that this is the start of your clean energy journey and ask you to remember that neither the destination nor the route you take are important – it's the impact of what you put in your tank that really matters!

Chapter 2

History of Diesel and the diesel engine

Early life

Rudolf Christian Karl Diesel was born on March 18, 1858 in Paris to Bavarian parents, Theodor and Elise Diesel, the second of their three children. Theodor had immigrated to France from Germany around 1850, he had a small leather working business there and provided a meager income for his family. Rudolf, a shy but bright child, spent most of his childhood in France. He exhibited an aptitude in mechanics from an early age but also excelled in mathematics and languages (he spoke three; German, French, and English).

During the Franco-Prussian War of 1870/71 Germans became political undesirables in Paris and so the family left for London. Theodor found it very difficult to obtain work there and

Figure 2.1
Rudolf Diesel.

when an uncle back in Augsburg, Germany, offered to take the now twelve-year-old Rudolf his parents agreed. Rudolf was entered into a three-year program at a technical college in Augsburg where he again excelled. After his parents returned to Paris in 1871, Rudolf stayed in Germany where he graduated as top of his class. He returned to Paris but after the tragic death of his elder sister returned again to Augsburg where he enrolled on a mechanical engineering program and again graduated top of his class.

Rudolf was awarded a scholarship to attend the Munich Institute of Technology where he studied thermodynamics under Professor Carl von Linde, the inventor of the ammonia refrigeration machine and the man who devised the first practical method of liquefying air. It was here that Diesel began to think about his ideas for a "heat engine."

At the time the dominant source of mechanical power was the steam engine; however, this suffered from very poor efficiency, between 6 and 10%, with less than one-tenth of the energy available to do useful work relative to the energy in fuel it was fed. Diesel asked himself if the heat could not be turned directly into mechanical energy rather than via the steam.

Three factors motivated him from this point on: mechanics, thermodynamics, and a social conscience.

Diesel graduated from Munich Institute of Technology in 1880 and became an apprentice pattern maker and engineer in Winterthur, Switzerland. He was quickly transferred to Paris, where his old teacher Carl von Linde arranged a job for him in the city's first ice-making factory, where he was quickly promoted to plant manager.

Here in Paris he filed for his first patent (on the "production of ice in glass containers"), became a connoisseur of the arts, a linguist, and a social theorist. He married Martha Flasche in 1883 (they had three children – Eugene, Heldi, and Rudolf). Carl von Linde arranged a franchise for Diesel to sell Linde's refrigerators in Southern Europe and the young family moved to Berlin.

Birth of the diesel engine

In 1892 from Berlin, Diesel filed for a patent at the Imperial Patent Office in Germany and within a year he was granted Patent No. 67207 for his proposed engine (Figure 2.2). This is also when he wrote his paper, "Theory and construction of a rational heat engine to replace the steam engine and contemporary combustion engine" which described an engine that could burn any fuel ignited not by a spark but by the temperature caused by the compression of gasses, a similar concept to refrigeration.

At this time the engine was only a concept: Diesel needed to build a prototype. With the help of the Augsburg Machine Works, among others, he produced his first working prototype in February 1894 in exchange for sales rights to most of Germany. Diesel had enormous technical problems to overcome. He continued to build prototypes, modify designs and

Figure 2.2
Rudolf Diesel's patent for the diesel engine.

experiment with fuels such as gasoline, kerosene, and lighting gas for several years. On December 31, 1896, Diesel ran the first engine he considered to be a success.

As word of his new invention spread, he started selling licenses to build and refine the design across the world. He sold the American patent rights to the brewer Adolphus Busch for one million marks. As companies bought licenses, Diesel quickly became a very wealthy man. He established a company, the General Society for Diesel Engines, in 1898 to manage the business; it bought the rights to his engine and assumed full control, paying Diesel 3.5 million German marks for this (a massive amount of money).

All was not rosy, however; years of constant work were having a serious effect on Diesel's health; he suffered severe headaches, exhaustion, and gout and was sent to various sanatoriums for rest. He suffered exhausting patent disputes, the Diesel engine not being the first to employ the principle of compression ignition. In addition, while Diesel had been very successful in his own business, he had also made some terrible investments and lost millions in other peoples ventures.

At the Paris Exposition of 1900 a diesel engine was exhibited and won the Grand Prize. Myth has it that it was exhibited by Diesel himself, but in reality at the request of the French Government it was exhibited by the Otto Company; however, it is true that it ran wholly on peanut oil. To quote Diesel himself:

> ... at the Paris Exhibition in 1900 there was shown by the Otto Company a small Diesel engine, which, at the request of the French Government, ran on Arachide (earth-nut or pea-nut) oil, and worked so smoothly that only very few people were aware of it. The engine was constructed for using mineral oil, and was then worked on vegetable oil without any alterations being made.

This is often cited as evidence of Diesel's foresight on biofuels but it seems that it was more to do with the French Government's desire to enable their African colonies to become more self-sufficient and less reliant on imported fuel. Diesel, again:

> The French Government at the time thought of testing the applicability to power production of the Arachide, or earth-nut, which grows in considerable quantities in their African colonies, and which can be easily cultivated there, because in this way the colonies could be supplied with power and industry from their own resources, without being compelled to buy and import coal or liquid fuel.

However, it is true that Diesel was a strong advocate of the use of biofuels, if only for his original desire for the engine to be operational almost anywhere on almost any fuel. In 1911 he said:

> The diesel engine can be fed with vegetable oils and would help considerably in the development of agriculture of the countries which use it.

And in 1912:

> The use of vegetable oils for engine fuels may seem insignificant today but such oils may become, in the course of time, as important as petroleum and the coal-tar products of the present time ... motive power can still be produced from the heat of the sun ... even when the natural stores of solid and liquid fuels are completely exhausted.

Death

Despite the publishing of another book in 1912, *Die Enstebung des Dieslmotors*, recounting the history of his invention, Diesel's problems continued to rise. He seemed to be sinking into a deeper depression over his financial difficulties. On the evening of September 29, 1913, he boarded the SS Dresden to cross the English Channel from Belgium to England to attend a meeting of the Directors of the British Diesel Company. He boarded the ship with three fellow directors. The three men took dinner on board and later strolled on deck before retiring to their cabins around 10 p.m., arranging to continue their conversation in the morning – Diesel was later described as "cheery and buoyant." When, as instructed, he was called at 6:15 a.m. the next morning he could not be found; his bed had not been slept in and his night shirt was laid out ready and his watch was next to the bed. Several thorough searches were carried out, but he was not to be found; he had utterly vanished.

The British, European, and U.S. press had a field day, suggesting that Diesel had been murdered by big oil business, the British Secret Service, the Germans, or the French.

On October 10, 1913, his body was found in the water off the Dutch coast. Following the custom of the time, personal effects were removed from the decomposing body and it was returned to the sea. Diesel's son identified the effects as belonging to his father.

Considerable speculation has grown around Diesel's death: No autopsy or official investigation was ever carried out, he had marked an X by that date in his diary; he left no suicide note and no will. However, he did leave an overnight valise with his wife, with instruction not to open it until the following week; it contained 21,000 German marks in cash and bank statements showing that they were nearly broke.

Rudolf Diesel has no known grave.

Legacy

After Diesel's death, the diesel engine underwent considerable development, and became a very important replacement for the steam engine in many applications. The diesel engine has undergone many refinements and modifications since those early years – the system of fuel injection has become much more sophisticated, and electronic controls now meter fuel more

Figure 2.3
German commemorative stamp – 100 years of the diesel engine.

precisely, achieving smooth running and low emissions while returning fantastic figures for economy. In some respects, the modern diesel engine is virtually unrecognizable from its humble beginnings, while in many ways, the fundamental principles, motivations, and ability to run on biofuels are as relevant today as they were when the idea of Diesel's engine was conceived.

Chapter 3

How does a diesel engine work?

It is not essential to have an understanding of diesel engine technology in order to successfully run your car on biofuels but, especially if you encounter any problems, a smattering of knowledge will help immensely when trying to find faults and troubleshoot problems.

Modern diesel engines are immensely sophisticated devices. Whereas a diesel engine "can" be one of the simplest internal combustion engines, with the ability for a diesel engine with a mechanical fuel injection pump to operate with no electrical input at all, the modern diesel engine has a large number of sensors and controls to ensure that you are consuming the most frugal amount of fuel, while producing the greatest power with the lowest emissions.

Figures 3.1a and b give some scale of the mechanical complexity of the modern diesel engine. For those of you who are relatively familiar with gasoline engines, much that is shown in these two pictures will seem familiar.

Differences between petrol and diesel engines

We are assuming that you have a familiarity with internal combustion engine technology. A diesel engine shares many common concepts with what you might call a gasoline engine, more properly called a "spark ignition engine." The fundamental notion of a piston being propelled by a burning mixture of fuel and air in a linear motion, with this motion converted into a rotary motion by a crankshaft, is common to both gasoline and diesel engines (see Figure 3.2).

Diesel engines differ fundamentally from the "spark ignition" gasoline engines that we are more familiar with, in the respect that there is no "spark plug" to ignite the mixture – instead, the diesel engine relies on the heat generated by compressing the charge of air and diesel. If you have ever noticed how the end of a bicycle pump heats up when you use it, then you will know what we are talking about; compressed air gets hot and the more you compress it the hotter it gets. How much the air/fuel mixture gets compressed is known as the compression ratio; it is much higher for a compression ignition engine than for a spark ignition engine.

The compression ratio is the difference in volume between the engine cylinder when the piston is at the bottom of its stroke, and the top of its stroke. You may see the initials "BDC" and "TDC" a lot when reading diesel engine literature. "BDC" refers to bottom dead center, which means when the piston is at the bottom of its stroke, and the connecting rod is perfectly central, whereas TDC refers to the converse, when the piston is at the top of its stroke.

Adiabatic heating

"Adiabatic process" is a term used in thermodynamics to describe a closed system where no heat enters or leaves the system. (The opposite to an adiabatic process, is an isothermal process where heat is free to enter and leave the system.) In a closed engine cylinder, in an "ideal" world we can consider this a closed system (of course the real world is not ideal, but sometimes engineers and physicists need to pretend that it is). The diesel and air is heated "adiabatically" by the rapid change in pressure caused by the piston coming up inside the cylinder. Another way of thinking about this is that there is a fixed amount of "energy" in the system; this energy is the movement of molecules of air and the mist of fine diesel droplets. By compressing this mixture there is the same amount of energy present, but in a smaller space. As pressure and temperature are related, as the pressure increases, so does the temperature. If you want another example, think about a bicycle tire pump. As you compress the air in the cylinder to pump up your tires, the cylinder wall begins to feel warm. The heat generated by the compression of the charge, is what is used to ignite the mixture in the cylinder, as opposed to a "spark plug" in a spark ignition engine.

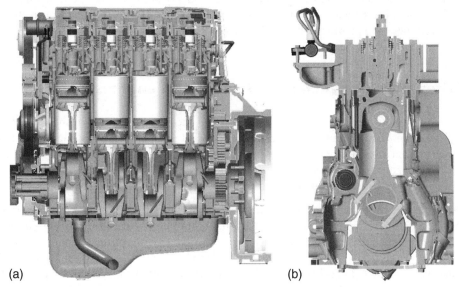

(a) (b)

▲ Figure 3.1
   (a) Cutaway common rail diesel engine (longitudinal cross section). (Courtesy of Cummins Ltd.)
   (b) Cutaway common rail diesel engine (transverse cross section). (Courtesy of Cummins Ltd.)

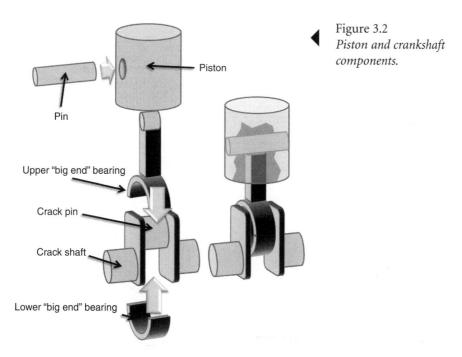

Figure 3.2
*Piston and crankshaft
components.*

Piston

Pin

Upper "big end" bearing

Crack pin

Crack shaft

Lower "big end" bearing

Furthermore, unlike a gasoline engine, which sucks a mixture of air and fuel into the cylinder (with the fuel being dispensed either by fuel injection into the manifold, or a more traditional carburetor), a diesel engine sucks only air into the cylinder, injecting the diesel fuel directly into the cylinder once the inlet valve has closed, as discussed later in this chapter when looking at diesel engine cycles.

If you want to do some further reading, the following web pages on diesel engine technology will reinforce and complement the information in this chapter:

auto.howstuffworks.com/diesel.htm.
www1.eere.energy.gov/vehiclesandfuels/pdfs/basics/jtb_diesel_engine.pdf.

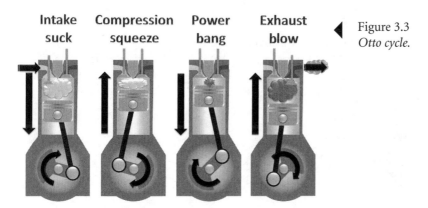

| Intake suck | Compression squeeze | Power bang | Exhaust blow |

Figure 3.3
Otto cycle.

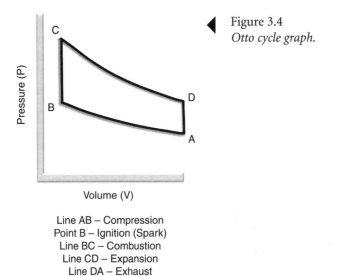

Figure 3.4
Otto cycle graph.

Line AB – Compression
Point B – Ignition (Spark)
Line BC – Combustion
Line CD – Expansion
Line DA – Exhaust

The diesel engine fuel system

In modifying a diesel vehicle to run on vegetable oils, we will be concerned with two of the diesel engine systems: the cooling system and the fuel system. Let's take a look at how diesel engines are fueled.

Diesel engines' fuel systems are more complicated than the simple fueling systems used for gasoline engines.

Diesel fuel is stored in a tank, usually at the rear of the vehicle, mounted below the trunk or below the rear seats. The fuel tank is connected first to the fuel filter by a pipe. Fuel filters can take different forms and there is some variation.

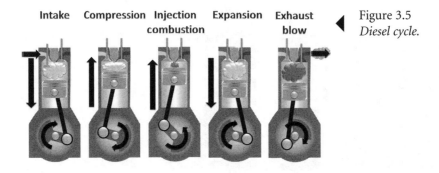

Figure 3.5
Diesel cycle.

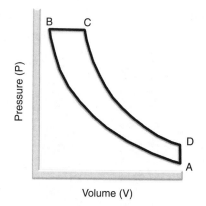

Line AB – Compression
Point B – Injection
Line BC – Combustion*
Line CD – Expansion
Line DA – Exhaust**
*Occurs at constant pressure
**Occurs at constant volume

▲ Figure 3.6
Diesel cycle graph.

Fuel injection

The technology of fuel injection is fundamental to a diesel engine's operation. Unlike gasoline engines, where fuel injection is a design option and a carburetor can be employed to meter the fuel/air mixture, with diesel, fuel injection is mandatory. This affects the way that the speed of the engine is regulated, as well as a host of other subtle things about the way the fuel is handled. This section should serve as an introduction to fuel injection technology; however, you may need to supplement it with further reading on information specific to your engine.

Direct/indirect injection

In some diesel engines, especially older engines, rather than being injected directly into the cylinder, the fuel is injected into a precombustion chamber, also known as a "swirl chamber," which is connected to the main engine cylinder. This chamber is designed to induce "swirl" into the fuel air mixture, ensuring effective mixing of the fuel and air, which aids with ignition.

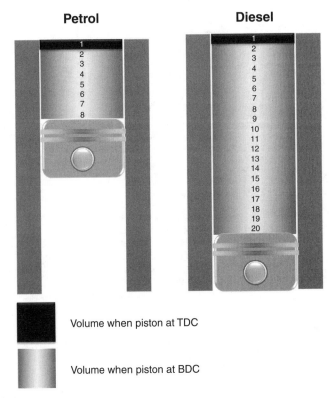

Petrol

Diesel

Volume when piston at TDC

Volume when piston at BDC

▲ Figure 3.7
Compression ratio explained BDC, bottom dead center; TDC, top dead center.

Fuel injection system configurations

Mechanically regulated fuel injector pumps

In-line injector pumps (port helix pumps)

An in-line injector pump consists of a series of "small pistons" in a row – hence the name in-line injector pump – with each individual piston supplying high pressure diesel for a fuel injector.

An in-line injector pump regulates the amount of fuel delivered to the injector, mechanically using a device called a port-helix valve, which can be seen in Figures 3.12 and 3.13.

The valve consists of a piston, which slides within a cylinder. The cylinder has a "port" in its side. The piston, which runs inside the cylinder, has a sloping groove cut around the outside.

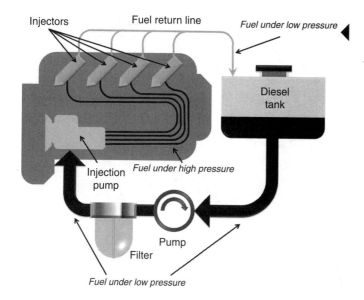

Figure 3.8
*Basic generic diesel engine
fuel system.*

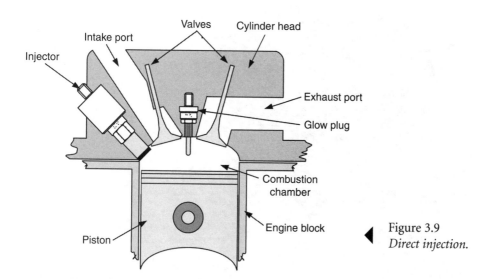

Figure 3.9
Direct injection.

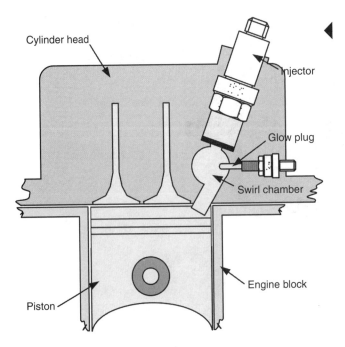

Cylinder head

Injector

Glow plug

Swirl chamber

Engine block

Piston

Figure 3.10
Indirect injection.

Figure 3.11
*In-line injector pump from
a large diesel engine.*

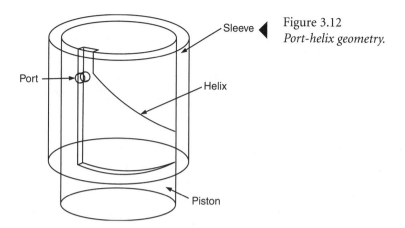

Figure 3.12
Port-helix geometry.

Think about how a slide runs around and round a helter-skelter. This provides the "helix" bit of the port-helix name.

The piston can move inside the cylinder in two directions, in and out, driven by a cam; this provides the "compression" of the diesel fuel.

Now, pay attention: This is the clever bit. The "port" in the side of the valve allows "excess" fuel to escape. The helix pattern allows the pump to dispense a different amount of fuel, depending on the "rotary" position of the valve in the cylinder.

The valves for all the cylinders are usually coupled together by some sort of rack and pinion that allows them to move together so that the same amount of diesel is delivered to all the cylinders.

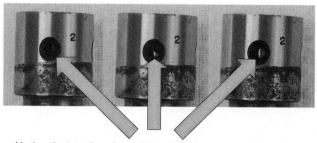

Notice that as the piston is turned, the port is progressively covered by the helix cut into the piston

Figure 3.13
Port-helix valve.

The cylinder is "under pressure" for all the time that the port is "covered up" by the solid metal of the cylinder. As the piston reaches the point where the "helix groove" uncovers the hole in the side of the cylinder, the pressure is lost.

By rotating the piston, the port can be uncovered sooner or later in the piston stroke, resulting in a larger or smaller amount of diesel fuel being dispensed by the injection pump.

Rotary injector pumps

In the same way that a distributor in a gasoline engine takes a rotary drive and turns it into a series of (electrical) outputs to produce a spark in each cylinder at the correct time, so a rotary diesel injector pump takes rotary motion, and turns it into the correct outputs of high-pressure diesel to supply the injectors in each cylinder at the correct time. Anecdotal evidence suggests that diesel injection pumps by Bosch, Denso, and many Japanese suppliers are more tolerant of the viscous fuels like SVO and WVO than Lucas and CAV injection pumps.

Mechanical unit injectors

Some older diesel engines use "mechanical" unit injectors, where the injection pump and the nozzle are combined into a single injector unit, which is actuated by a camshaft. This system does away with the complexity of a seperate mechanical injector pump, and removes the need for high-pressure diesel pipework, fuel being supplied to the injectors at low pressure.

Electronically actuated fuel injection systems

Common rail injection. In many systems the fuel injection pump serves the purpose of both delivering fuel at a suitably high pressure and providing synchronization with the engine timing, delivering the fuel to the right place at the right moment mechanically.

In common rail systems, the injector pump merely pressurizes the fuel that is supplied to all the injectors on a common rail; the injectors' actuations are controlled and synchronized to the engine electronically.

Some common diesel engine acronyms
CDI, common rail diesel injection
HDI, high-pressure direct injection
UTD, uniJet turbo diesel (Fiat group's common rail diesel technology)
HEUI, hydraulically actuated unit injection
SDI, suction diesel injection
TDI, turbo direct injection
TD, turbo diesel

Fuel injectors explained

A fuel injector consists of a number of components: namely, a nozzle body and a needle. We can see an exploded fuel injector in Figure 3.14, to help us understand the consituent parts of the injector. There is a bore running through the nozzle body, in which the needle can move freely. The needle acts as a seal against the high injection pressures. The needle has a "seat," and the injector body has a conical recess into which the injector needle's "seat" fixes.

A spring which bears down on the needle maintains the integrity of the combustion chamber – only high-pressure fuel from the injection pump can move the needle to allow fuel to be injected. When the pressure of fuel between the needle and the seat exceeds the pressure of the spring bearing down on the needle, the needle lifts and fuel is dispensed. This is often referred to as the cracking or crack pressure.

Because the orifice through which the fuel is administered to the cylinder is small, the high-pressure fuel expands rapidly and atomizes, mixing thoroughly with the air. At the end of the injection period, when the fuel pressure in the pipe from the injection pump drops below a certain level, the needle closes, sealing the compression chamber.

There is a subtle difference between the way that the nozzle is shaped after the injector needle, which leads to a differentiation between the two different types of injectors. You will remember back to the concept of "direct" and "indirect" injection: Well, the different injectors suit different applications. In the first injector, shown to the left of Figure 3.15, and shown in close-up in Figure 3.16a, we have the "throttling pintle" injector nozzle, which is commonly found in "indirect" injection engines, where the fuel is first sprayed into a precombustion chamber. Pintle nozzles produce a "cone-shaped" spray pattern. This is a characteristic of its nozzle design, where the needle, which has a small "pintle" on the end, protrudes through a hole at the end of the injector.

By contrast, hole-type injectors, which are more commonly used in direct injection engines, have an additional piece of metal at the end of the injector, shrouding the end of the needle, with a number of holes offset at an angle. With the nozzles located below the edge of the seat, the fuel must pass past the needle into a small cavity, before spraying through the holes, which produce a number of smaller plumes of fuel, which are injected directly into the cylinder. We can see that with a hole-type fuel injector, as with the pintle nozzle, the high pressure of the fuel inside the injector allows the fuel to atomize as it is sprayed into the cylinder, ensuring that it mixes well with the air.

▲ Figure 3.14
Component parts of fuel injector.

Pintle type
nozzle

Hole type
nozzle

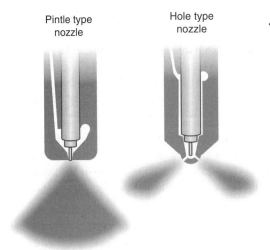

Figure 3.15
Fuel injector nozzles compared.

(a)

(b)

Figure 3.16
(a) Close-up of the pintle nozzle injector.
(b) Close-up of the hole nozzle injector.

▲   Figure 3.17
Fuel injector sprays a fine mist.

Turbochargers

A turbocharger allows a greater amount of air to be squeezed into a cylinder. It uses some of the energy in the escaping exhaust to spin a small turbine, which, in turn, spins another small turbine that is used to push air into the cylinders – more air than would have got in there otherwise – which results in greater efficiency. The more air and diesel (in the correct ratio) that you can squeeze into a given space, the more power that you can extract from the combustion of the fuel. However, at "atmospheric pressure," which is the pressure at which "normally aspirated" engines are filled, there is a given amount of air that will fill a given volume.

Diesel fuel and air burn in a given ratio, called the stochiometric ratio – put in too much fuel, and the mixture is rich, not enough fuel and the mixture is lean.

Helping the engine to start – glow plugs

It's worth mentioning glow plugs, as they are often a point of confusion when trying to understand how a diesel engine works. A glow plug does not fulfill the same function as a

Figure 3.18
Testing fuel injector spray patterns.

Figure 3.19
Pattern produced by a multi-hole fuel injector.

spark plug in a spark ignition engine: It does not ignite the fuel like a spark plug – glow plugs are only there to assist with cold-starting.

It's like this … as we have discussed earlier in the chapter: When you squeeze a mixture of fuel and air into the engine's cylinder, it gets hotter because it is a closed thermodynamic system and the pressure is increasing. However, the fuel and air need to reach a certain temperature for ignition to take place, and when the engine is cold a lot of the heat from the compression is going to escape into the cold engine block. This causes bad ignition and the engine will run roughly, if it runs at all.

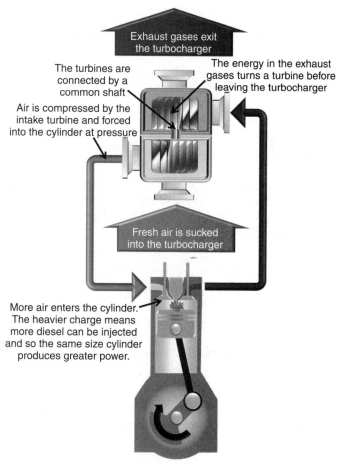

▲ Figure 3.20
How a turbocharger works.

▲ Figure 3.21
A glow plug.

The idea of including a glow plug is to preheat the cylinder so that smooth combustion can take place inside the cylinder. Referring to Figure 3.22, we can see that a glow plug is simply an electrical heating element that is "ruggedized" and designed for the harsh environment of the diesel engine. The engine block provides an electrical ground return, while a supply is delivered to the terminal of the glow plug.

Improving the efficiency of your diesel engine

The following modifications are not for the faint hearted; however, it is possible to make some fairly simple modifications to your diesel engine to improve efficiency. If you are not confident with these procedures, seek professional assistance – a good engine shop may be able to carry out the work for you for much less than a damaged cylinder head would cost to replace!

Diesel engines are mass-produced items and, as with any commercial item, there is a balance to be struck between the quality of the product and the price it retails at. The more work that goes into a casting by an engine manufacturer, the greater the cost of the final product, so that manufacturers have to trade off a compromise between the amount of work they are able to carry out on the casting and its performance. Thankfully, much of this extra finishing work can be carried out afterwards; however, it will entail a strip down of your engine and some serious work!

The work you can do on your cylinder head is focused around smoothing the path of the airflow to enable it to flow through the cylinder head while encountering the least resistance as possible. There are several options for improving your cylinder head.

Removing obstructions

Sometimes, castings are made with "extra metal" where it is helpful for strength/durability of the engine, cost-effectiveness of the casting, or is just sloppy design. It is sometimes possible

Terminal

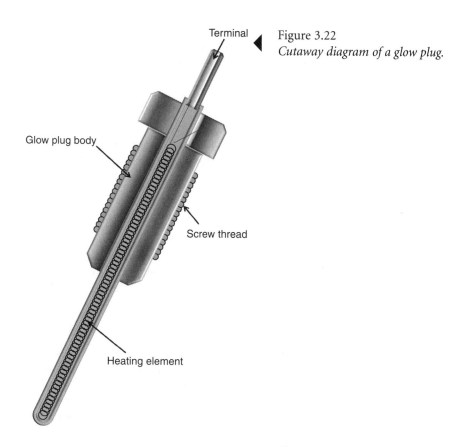

Figure 3.22
Cutaway diagram of a glow plug.

Glow plug body

Screw thread

Heating element

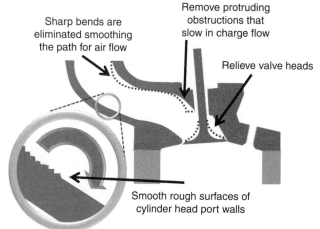

Sharp bends are
eliminated smoothing
the path for air flow

Remove protruding
obstructions that
slow in charge flow

Relieve valve heads

Smooth rough surfaces of
cylinder head port walls

Figure 3.23
Efficiency modifications to your cylinder head.

that there is extra metal, e.g., in valve guides, which obstruct the airflow through the cylinder, and could be smoothed or removed to optimize the airflow. Removing obstructions allows the air to flow more freely and optimizes the performance and efficiency of your engine.

Eliminating sharp bends

Sometimes, for practical reasons, the bends in the air's path between the inlet manifold and the inlet valve are sharper than optimal. It is possible in some instances to carefully remove metal in the path inside the cylinder head, to allow the air a smoother flow.

Smoothing cylinder walls

Castings come with a less than perfect surface finish, and often, in the name of cost, castings have a less than smooth internal surface finish in the air path through the cylinder head.

Relieving valve stems

It is possible to remove a small quantity of metal from the stem of the inlet and exhaust valves to allow the intake charge to flow into the cylinder more easily and the exhaust charge to exit the cylinder as quickly as possible. The less metal in the path of the gases flowing into and out of the cylinder, the quicker they can flow. You want to leave a smooth surface; remove sufficient metal to smooth the flow of gases, but do not compromise the integrity of the valve.

Chapter 4

Collecting waste vegetable oil

Waste vegetable oil (WVO), or used cooking oil (UCO), or used vegetable oil (UVO) or whatever you prefer to call it is the stuff that is used for deep-frying by restaurants, pubs, burger vans, and fish and chip shops everywhere. It is the raw ingredient for biodiesel home brewers and WVO enthusiasts. This short chapter is all about how to get hold of it and how to select the good stuff.

Collecting WVO is common to Chapter 6 and Chapter 9, so rather than repeating ourselves too much, we have put the information into this handy little chapter instead.

> For the perils of overenthusiastic collection of WVO, check out Issue 5 of "BiodieselSMARTER," which chronicles how one unfortunate biodieseler had to dispose of over 3,500 gallons of WVO. After setting up happy liaisons with local purveyors of grease, the oil kept flowing in, piling up in the back yard; however, processing ability lagged behind, causing an "oil mountain." Moral: Only collect 500 gallons a week of oil if you have the capacity to process it: Save your enthusiasm for a day when your plant and processes are up the job!

Buying WVO

It is perfectly possible to buy waste oil too, but harder to control its quality. Find out who your local waste oil collectors are and see if you can strike a deal with them.

Choosing the good stuff

Its quality varies massively from one place to another. Chinese restaurants tend to have the best stuff, but it varies massively. Good-quality oil will tend to be clearer and lighter in color; bad-quality oil will often be thicker or darker or creamy due to overheating or overuse.

Take your time, finding a good source of oil will make your biofuel-life easier in the long run.

Good WVO has not been used too much or been overheated; it will have a low acid number (see How to titrate, p. 47) and will not have much, if any, water in it (see How to test for water in Chapter 6).

If its acid number is more than 3 or 4, you may not want to use the WVO for making biodiesel, at least until you are more experienced at making it; if it is greater than 10, it is useless for biodiesel unless you are using the acid–base method. If you are thinking of using it in a converted engine, rather than for making biodiesel, take a look at Chapter 9 but as a rule of thumb oil titrating at higher than 3 or 4 is often considered too acidic.

Collecting it

How you go about finding, negotiating, and collecting your raw ingredient, WVO, is very important; do it wrong, and you could alienate a restaurant manager or even find yourself in trouble with the law.

You should not simply take the stuff you find behind the restaurant. Just because we are calling it waste, does not mean it is without value. It almost certainly belongs to someone (the oil collector, the restaurant, another biofuel maker), and it is very likely the restaurant won't take kindly to you helping yourself without permission.

In the UK, pubs and chip shops usually pour the used oil back into the 25 liter drums or the "cubies" (cubelike plastic containers in a supportive cardboard box) it arrived in to

Figure 4.1
Waste fish and chip shop oil in Wales.

await collection. In the United States, it is more normal for the oil collection company to provide a dumpster for the used oil.

Start by going into the place as a customer; pick a time when they are not too busy. Try to look neat and tidy, maybe order something, and ask to speak with the manager.

Explain to the manager, very briefly, that you make biodiesel, even if you are not planning on actually making biodiesel, from waste vegetable oil, and ask if you could take a little of their waste oil to test its suitability.

In years gone by, biodieselers had trouble convincing restaurant managers why they really wanted the oil (making fuel from it was just too improbable), but the chances are that everyone has heard of biodiesel today. However, this is not the time to be explaining the subtleties of SVO and WVO and biodiesel and blending – keep it simple.

At first explain that it is only an experiment at this stage: You need only a small amount, maybe a gallon. You don't want to be making a deal with someone who is going to give you bad oil, which makes you have to back out of it, alienating managers and making life harder for a biofuel maker who follows you.

If the manager agrees, then take your sample home and test it; if it is good, go back and strike a deal.

Regulations

Regulations vary massively from country to country and are constantly changing, so check the situation in your local area.

In the United States the oil, once poured in, often becomes the property of the dumpster owner, so providing another container for your oil may help make you legal in this situation.

The law in the UK regards used cooking oil as an industrial waste and its storage, transport, and disposal are regulated accordingly by the Environment Agency (EA), regardless of whether you are a private individual, club, or commercial business. In order for you to legally collect the waste oil, you need to be a registered waste carrier; transporting the oil is illegal if you are not registered. The cost is about £150 (US$300) for 3 years and the application is pretty much just a formality; send off the money and the form and you will be issued a Waste Carriers License. Being a registered waste carrier will not only enable you to move the WVO legally but also you will be able to issue Waste Transfer Notes: Something pubs, restaurants, and schools should want to receive from you in return for their WVO in order to prove they disposed of it properly, but in reality a lot don't seem to care. The issuing of licenses and Waste Transfer Notes enables the EA to keep track of where waste is being produced, who is moving it, and where they are moving it to. This may initially seem over-the-top, but the rules are there to protect the environment from unscrupulous people who would dispose of waste inappropriately.

More information about waste carrying, Waste Transfer Notes, and the Duty of Care can be found on the EA's website: www.environment-agency.gov.uk.

The caterer you are looking for to collect the WVO from may have already paid the cost of disposal in the price he paid when buying the oil; it is often the same company who supplies, delivers, collects, and disposes of the oil. However, many catering companies in the UK don't buy their oil from a dedicated oil supplier and have to find ways of disposing of their WVO themselves; these methods are sometimes not entirely legal and sometimes involve pouring the oil down the drain. While maybe not entirely legal, you should also not beat yourself up too much if you find yourself collecting WVO from these guys; you are doing them, the sewage company, and the environment a big favor.

Transporting it

Cubies are easy to load into your car, but can still be dirty and oily and the empties present a disposal problem for you afterwards.

Once you have loaded up the containers of oil into your vehicle, it is well worth triple-checking that they are secure; the last thing you want is for them to fall over and spill in your car. Eugh! Also, don't overdo it; you don't want to be overloading your car with oil just because it is there. It is dangerous and illegal, so make two runs instead.

Other caterers may have their oil in a bigger container requiring you to pump it from their container into yours. Rotary pumps are good, simple, and reliable; 12V electric pumps can also work great. Any big container in your trunk to transfer the WVO into will do just so long as it's secure: not too heavy when full of oil and not going to spill everywhere on the trip home.

Suckers

An old refrigerator's compressor makes a great vacuum pump; you can build a vacuum-proof container into your vehicle or a trailer, maybe from an old gas tank, suck all the air out of it when still at home, and then drive to your favorite burger joint and suck up all their WVO.

 You can buy plans for a sucker at Murphys Machines: www.murphysmachines.com/supersucker.html

Storage of waste oil

In the UK, storage of waste oil is regulated under the Waste Management Licensing regulations. However, they do allow you to keep up to 100 liters of waste cooking oil for the production of fuel without worrying about licenses. The waste cooking oil and the biodiesel must be securely stored and a "bund" is always a good idea, often a legal requirement.

Again, the regulations are updated often and vary from one place to another.

Health and safety

This is often dirty work, so you will need to make sure you are not wearing your favorite jeans and have some gloves on. Good shoes with good grip are a good idea, as you will be working in some slippery places, and steel toecaps if you are handling anything heavy. It is a good idea to have a change of shoes with you, as driving with oil on the soles of your boots could be dangerous.

How to titrate

What is titration?

Titration is a quantitative chemical analysis used to determine the concentration of a substance. Sounds complicated and scary? Well don't worry, it is actually very easy. As far as we biofuels people are concerned, titration is a simple test we perform on used oil (WVO) in order to determine how strong its free fatty acid (FFA) concentration is.

Why titrate?

Unlike new, unused oil, your WVO has FFAs in it. FFAs are made when the fatty acid "legs" of the triglyceride oil molecule get broken off. The more the oil is used and the hotter it gets, the more legs get broken off and the higher the FFA concentration.

When we make biodiesel, we need to know how much of the FFA stuff there is in our WVO in order to know how much extra catalyst to add.

When using the WVO in an SVO converted engine, we need to know how strong the FFA concentration is because we want to manage the risk of possible damage to our engine.

Requirements

- Some small jars or beakers
- An empty bottle to store your testing solution in

- A marker pen or sharpee to write on it
- Chemical resistant gloves
- An apron
- Some goggles
- Some lye, catalyst, KOH or NaOH
- Distilled water; ordinary water is no good
- Isopropyl alcohol, rubbing alcohol
- An indicator: Phenol red works, as does, turmeric or red cabbage
- Some small syringes, a few small sizes, the kind you get a shot with, but without the needle. Some pharmacies may be willing to help you, or maybe your local veterinarian
- A way of accurately measuring 1 liter of the distilled water
- A way of accurately measuring your catalyst

How to do it

Basically, we are reacting small amounts of our chosen catalyst (lye) with the FFAs in our oil and measure the pH as we proceed. The normal way of doing this is to have an indicator in the reaction. The indicator will change color when we reach the pH we are looking for and, because we were paying attention to how much catalyst we have added, we know how much it took to neutralize the FFAs.

The test will give us a number, called the *acid number*.

Step 1

First, we need to make our reference testing solution, a solution of 0.1% NaOH; it's not as tricky as it sounds. Using our scales to measure the catalyst, simply dissolve exactly 1 g of it in exactly 1 liter of water. Improved accuracy will be achieved by dissolving X grams in X liters of water; this will give you a lot of reference solution but it keeps reasonably well so long as it is kept in a closed container and is not exposed to the air. Each milliliter of your distilled water now contains one-thousandth of a gram of catalyst.

If you are planning on making biodiesel from your WVO, you must always use the same catalyst for titrating as for making the biodiesel.

Now, label the bottle clearly so you know what it is next time and don't get it confused with some other bottle of clear liquid.

Step 2

We need to do a thing called a blank titration, i.e., we need to check to see the pH of our isopropyl alcohol: Usually it is neutral but sometimes it's not and if we don't check, it will give us an inaccurate result when we titrate the oil later. The procedure for the blank titration is exactly the same as for the oil, only without the oil.

Measure about 10 ml of isopropyl alcohol from a syringe (the one you wrote "isopropyl alcohol" on, so as not to get it confused with the others) into a small, clean dry jar or beaker; add to it a little of your indicator.

Now, using the testing solution syringe, add the testing solution to the alcohol drop by drop, swirling the beaker as you go, until it changes color. Stop when you have neutralized the acid in the alcohol; from now on we are only measuring the pH of the oil.

Next, test the WVO. Measure, using your WVO syringe, *exactly* 1 ml of WVO into your jar of alcohol. Swirl the mixture to keep the oil suspended in the alcohol.

Keep swirling and add the testing solution in tiny increments, keeping a note of exactly how much you have added, until the beaker's contents change color and stay that way for about 30 seconds. You have neutralized the FFAs.

It is easy to make a mistake doing this. Perform the test a few times and obtain a consistent result before proceeding.

So how many milliliters of testing solution did it take to neutralize the FFAs? Each milliliter of testing solution equals the number of grams of *extra* catalyst you will need to add to your biodiesel reaction. If it is more that 3 or 4, you may not want to use it for making biodiesel until you are more experienced at it; if it is >10, it is useless for biodiesel unless you are using the acid–base method. If you are thinking of using it in a converted engine, take a look at Chapter 9.

The Collaborative Biodiesel Site has an excellent article on how to titrate: http://www.biodieselcommunity.org/titratingoil/.

There are a number of excellent videos showing how to titrate on Youtube (www.youtube.com); search for how to titrate.

Testing for water

(This section is repeated in Chapter 9.)

Testing for water in your WVO is simple, if rather crude. Take a small cupful of your oil in a pan and gently heat it to around 100°C (212°F). If there is bubbling or spitting, you have water present. How much is usually a matter of guesswork and experience.

See the Dewatering section of Chapter 9 for advice on separating the water from the oil.

Quantitative test for water

You know you have water present, but how much? Calculating how much water is in your WVO is not difficult. Take a representative sample from your WVO by first ensuring it is well mixed. Take a few hundred milliliters, measure quite carefully, and weigh it accurately. Put it into a pan and heat it gently; let the water in it boil, but not so much as it spits out of the pan. Careful here: Hot oil is the number one cause of household fires and can give you nasty burns. When it has stopped spitting, turn off the heat, allow it to cool, and weigh it again. The difference between the two numbers is how much water you had; 1 ml of water weighs exactly 1 g, so if you know how many milliliters of wet oil you started with, you can calculate how wet the oil was as a percentage.

However a quantitative test is of limited use because, whether you're making biodiesel or using the WVO straight, anything more than the tiniest amount of water is not acceptable.

Chapter 5

Biodiesel chemistry

Introduction to the chemistry of biodiesel

What follows is a guide to the chemistry of biodiesel for the nonchemist. In this section, we are going to be introducing some chemistry terminology, and ways of visualizing the process, that may seem familiar from high-school chemistry. If this process seems to fly by a little fast, links have been included to web material that will help you brush up on your chemistry knowledge. If however, your chemistry skills are fairly sharp, you might want to skip the next few pages and look at the "Transesterification" reaction, which is where things will really start to get interesting.

None of this is particularly scary stuff, once you are familiar with the process. Later in this book, we will be exploring this in a more practical manner, which hopefully will help you make sense of some of the reactions that are presented in this chapter. You do, however, need to bear with the theory, as understanding of the theory is essential to being able to produce biodiesel successfully.

We're going to start by revising some really simple high-school chemistry terminology that you will need to be familiar with in order to understand the processes that take place when we produce biodiesel. Here we will be covering the chemical theory behind biodiesel. In later chapters, we are going to go through the practice, starting with small samples and working up into larger batches.

Chemistry refresher

If your knowledge of chemistry is a little rusty, you might want to take a look at online resources to brush up.

www.chemguide.co.uk/.

Definitions

Acid

An acid is a chemical compound which has a pH (measure of acidity/alkalinity) less than 7.0 (7 is considered neutral). Acids are corrosive. They have a hydrogen ion activity greater than water, which is considered neutral.

Base/alkali

A base/alkaline substance, is a chemical compound which has a pH greater than 7.0. Bases/alkaline substances are caustic. Bases/alkaline substances "accept protons" in chemical reactions. In the context of this book, we will be using a strong base as part of the methoxide mixture that converts our oil into biodiesel.

Catalyst

A catalyst facilitates the reaction between other substances, but is not consumed itself.

Emulsification

An emulsion is a mixture of two immiscible (means that they won't mix!) liquids. If you take a glass, half fill it with water, and top up with oil – you will find that they won't mix – the oil floats on top of the water. However, add a chemical known as an "emulsifier" and you'll find that the two fluids will mix together easily, with droplets of one liquid being suspended in the other liquid. This is great if you want to make a low-calorie spread, but not so grand if you want to make biodiesel.

A quick test for the presence of emulsifiers is to take a small container – half fill with water, and half fill with biodiesel. Cap off the container and shake it vigorously for a short period. Then leave the liquid to settle. If no emulsifiers are present, the oil should find its level on top of the water, settling out relatively quickly.

If however, your fluids don't settle into two distinct layers within a couple of minutes, there are still emulsifying agents present: These could be soaps or mono- and diglycerides; see Chapter 6 for more information on this.

Organic chemistry, hydrocarbons … and all that jazz!

The branch of chemistry that this chapter is primarily concerned with is "organic chemistry," which is principally concerned with the chemistry of compounds containing carbon and hydrogen.

Strictly speaking, a hydrocarbon contains only hydrogen and carbon atoms – we refer to compounds containing only hydrogen and carbon as "pure" hydrocarbons; although *strictly speaking* they are not hydrocarbons, compounds containing additions other than hydrogen and carbon are often referred to as "impure hydrocarbons."

Representing chemicals

In this chapter, we are going to be using a variety of different ways to represent chemicals in order to help you understand the nature of how they are composed. The two main notations that we will be using are the "structural" notation, "letters" joined by lines, with each line representing a bond between two different molecules, and "ball-and-stick" models, which help you get a better idea of what the chemicals look like in three-dimensional (3D) space.

Alkanes

Alkanes consist of carbons connected by a single bond with hydrogens, forming a single chain hydrocarbon. The word hydrocarbon implies that alkanes consist of hydrogen and carbon but nothing else. Alkanes are aliphatic: i.e., they consist of straight and branched chains only, as opposed to aromatic compounds which contain "rings" of carbon. We aren't going to deal with aromatic compounds in this discussion, so you can happily forget about them for the time being.

The alkane series are also referred to as the paraffin series.

First, we will take a look at methane. You will see we have two representations of methane. The first shows us structurally the configuration. We have four hydrogen atoms connected each with a single bond to a carbon atom. Methane is the simplest hydrocarbon around; it is straightforward, all single bonds, and represents a carbon atom with all its "binding sites" occupied with hydrogen atoms.

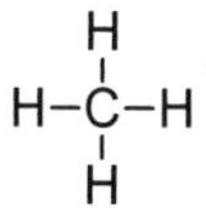

H−C−H with H above and H below

Figure 5.1
Structural representation of methane.

If we look at the ball-and-stick model of methane, we can begin to visualize the way that the atoms fill physical space.

Figure 5.2
Ball-and-stick model of methane.

Looking at the next largest chain in the alkane sequence, we turn to ethane. The fundamental difference between ethane and methane is that ethane has an additional carbon atom. This takes the place of one of the "hydrogen" atoms in the methane structure, connecting to the hydrogen by a single bond. We can see that this additional carbon comes with two hydrogen atoms of its own.

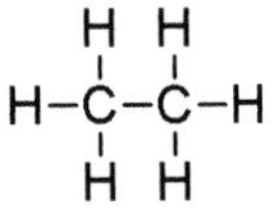

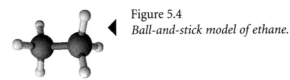

Figure 5.3
Structural representation of ethane.

Looking at the ball-and-stick model of ethane, we can see how the atoms are arranged in three dimensions (3D).

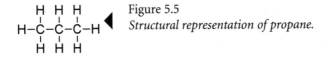

Figure 5.4
Ball-and-stick model of ethane.

The logical extension of this is propane, the gas we use in our flame grills and barbecues!

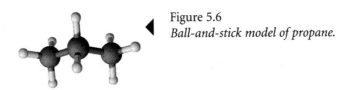

Figure 5.5
Structural representation of propane.

Figure 5.6
Ball-and-stick model of propane.

The general formula for alkanes is C_nH_{2n+2}.

Chemicals are named using a fairly structured numbering system that refers to the number of carbons contained in a chain. This is the system adopted by the International Union of Pure and Applied Chemistry, and it is simple to follow once grasped. We present here the prefixes up to 20, which will become apparent in our discussion of the chemistry needed to understand Biodiesel production (Table 5.1).

Alkenes

Alkenes are also hydrocarbons; however, they differ from the alkanes in that they possess a "carbon=carbon double bond." This bond is stronger than a single bond; however, where a

Table 5.1 Chemical prefixes

Prefix	Number
Meth	1
Eth	2
Pro	3
But	4
Pent	5
Hex	6
Hept	7
Oct	8
Non	9
Deca	10
Undeca	11
Dodeca	12
Trideca	13
Tetradeca	14
Pentadeca	15
Hexadeca	16
Heptadeca	17
Octadeca	18
Nonadeca	19
Icosa	20

For further details on the IUPAC numbering system beyond 20, a good resource to consult is http://en.wikipedia.org/wiki/IUPAC_numerical_multiplier.

double bond is formed, the "extra bond" must come from somewhere, so it takes the place of a hydrogen atom.

Ethene is the simplest alkene, consisting of two carbon atoms and two hydrogen atoms.

Figure 5.7
Structural representation of ethene.

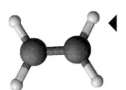

Figure 5.8
Ball-and-stick model of ethene.

Look what happens when we add an additional carbon and two hydrogens to produce propene. Look at how the atoms are configured.

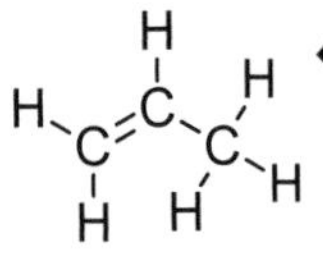

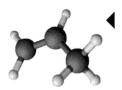

Figure 5.9
Structural representation of propene.

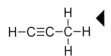

Figure 5.10
Ball-and-stick model of propene.

Alkynes

Next, we are going to look at a series of chemicals called the alkynes, sometimes referred to as the "acetylenes" or the "acetylene series."

Ethyne, the simplest alkyne, is also known as "acetylene": welders will be familiar with this compound!

Note that here, rather than a "double bond" we now have a "triple bond" – see that another hydrogen has been "sacrificed" to lend its bond to the triple bond.

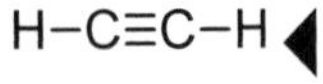

H−C≡C−H

Figure 5.11
Structural representation of ethyne.

Figure 5.12
Ball-and-stick model of ethyne.

Propyne is the next logical extension. See the pattern emerging?

H−C≡C−C−H

Figure 5.13
Structural representation of propyne.

Figure 5.14
Ball-and-stick model of propyne.

Introduction to functional group chemistry

One way that we can think of chemistry is in terms of "functional groups." There are certain "patterns" that we see time and time again in chemical compounds. We give certain patterns of chemicals that are joined in certain ways their own names, and we can think about them as a "group" that we can join in place of a hydrogen bond. Think back to our alkanes, alkenes, and alkynes – each time, we switched a "hydrogen–carbon" bond for a "carbon–carbon" bond. With functional groups, we can swap a hydrogen bond for a link to a functional group, or we can "break apart" a double bond to leave a single bond and two "connections," which we could then link to a hydrogen or a functional group. Thinking about chemistry in this way makes it easy to understand like a jigsaw puzzle.

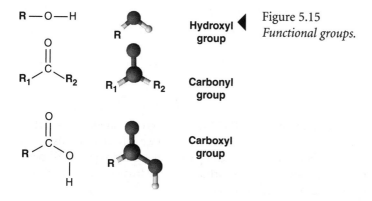

Figure 5.15
Functional groups.

Alcohols

Alcohol eh!… Now I have your attention!

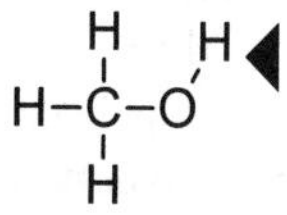

Figure 5.16
Structural representation of methanol.

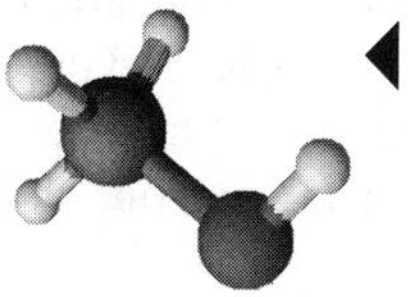

Figure 5.17
Ball-and-stick model of methanol.

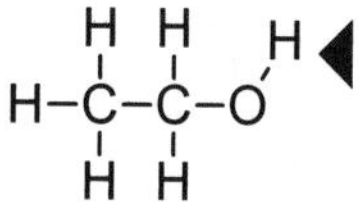

Figure 5.18
Structural representation of ethanol.

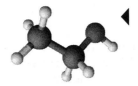

Figure 5.19
Ball-and-stick model of ethanol.

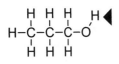

Figure 5.20
Structural representation of propanol.

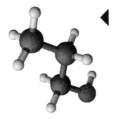

Figure 5.21
Ball-and-stick model of propanol.

An alcohol is a chemical compound that has an OH (hydroxyl) group bonded to a carbon atom. We can see the simplest acid "methanol" (Figure 5.17), and we can see how closely this resembles methane (Figure 5.1) apart from an —OH group where we would expect a hydrogen. We can also see how this chain grows by adding carbons and hydrogens, so ethanol is to ethane and propanol is to propane as methanol is to methane.

Fatty acids (carboxylic acids)

Fatty acids, also known as carboxylic acids, follow a similar pattern to the alcohols described above. Look at the image of methanoic acid (Figure 5.22), and try to compare it to the image of methane. From the four potential bonds that the carbon atom can make, one has been replaced by an —OH group (as in methanoic acid above); however, an additional two bonds are given over to a double bond with an oxygen atom, leaving only one hydrogen. We can "grow" fatty acids chains, just like with our other hydrocarbons, by adding units of carbon and hydrogen. So see how methanoic acid transforms to ethanoic acid (Figure 5.24) and then on to propanoic acid (Figure 5.26) with the addition of a handful of carbon and hydrogen atoms. We are interested in much longer fatty acid chains, the kind found in fats and oils that we can process to make biodiesel.

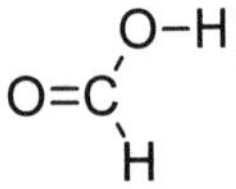

Figure 5.22
Structural representation of methanoic acid.

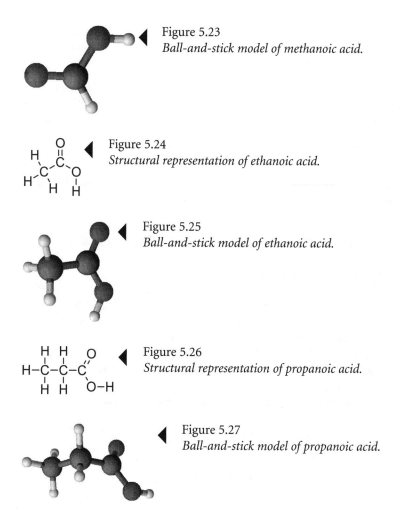

Figure 5.23
Ball-and-stick model of methanoic acid.

Figure 5.24
Structural representation of ethanoic acid.

Figure 5.25
Ball-and-stick model of ethanoic acid.

Figure 5.26
Structural representation of propanoic acid.

Figure 5.27
Ball-and-stick model of propanoic acid.

Oils and fats

Oils and fats belong to a group of biological chemicals called "lipids" – what distinguishes oils from fats is that oils are fluid at room temperature, whereas fats are solid. So it's all a question of the melting point: Heat a fat up, and it turns into an oil. Chemically, they share the same components; which is to say a "glycerol backbone" that joins together three chains of "fatty acids." So, structurally, fats and oils are identical, just that their melting point differs.

Our triglyceride chains contain mainly carbon and hydrogen atoms, with very few oxygen atoms (6 per triglyceride molecule). This means that chemically we can call them "very reduced," which means that the carbon and hydrogen atoms haven't yet been oxidized.

This is what makes fats and oils good as an energy storage medium – as in "oxidizing" the triglycerides – whether this be by chemical processes inside our body, or in an internal combustion engine, we can release a lot of energy.

Fats and oils are triesters of glycerol, propan-1,2,3-triol (more about esters later on). The "tri," which you should recognise from "triglyceride," means that there are three chains, and it is the esters that we are interested in, and that we want to harvest to make our biodiesel.

So, if, chemically, oils and fats share the same basic structure, what is it that makes them different and gives different types of vegetable oil different properties?

One of the factors that distinguises oils and fats is the length of the chain of hydrogen and carbon attached to the glycerol backbone.

Saturated, unsaturated, and polyunsaturated fats and oils

Another thing that differentiates different fats and oils is that some fats and oils have chains where all the bonds are single bonds, and the bonding sites of each carbon atom are all filled with hydrogen atoms, whereas some fats and oils have one or more double bonds along the length of the chain. This affects the structure of the molecule, as when a double bond occurs, the chain of hydrogens and carbons can "kink." These "kinks" in the chain mean that unsaturated and polyunsaturated chains do not stack particularly neatly together. This "lack of fit" means that the forces holding everything together are "stretched" (if you like by the awkwardly shaped chains), meaning that they can be separated more easily. This leads to lower melting points in unsaturated and polyunsaturated fats and oils than in saturated oils.

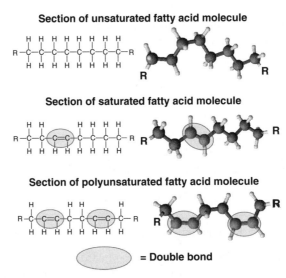

Figure 5.28
Saturated, unsaturated, and polyunsaturated fatty acid molecules.

Table 5.2 Common saturated fatty acids

Common name	Systematic name	Structural formula
Butyric	Butanoic acid	$CH_3(CH_2)_2COOH$ or C4:0
Caproic	Hexanoic acid	$CH_3(CH_2)_4COOH$ or C6:0
Caprylic	Octanoic acid	$CH_3(CH_2)_6COOH$ or C8:0
Capric	Decanoic acid	$CH_3(CH_2)_8COOH$ or C10:0
Lauric	Dodecanoic acid	$CH_3(CH_2)_{10}COOH$ or C12:0
Myristic	Tetradecanoic acid	$CH_3(CH_2)_{12}COOH$ or C14:0
Palmitic	Hexadecanoic acid	$CH_3(CH_2)_{14}COOH$ or C16:0
Stearic	Octadecanoic acid	$CH_3(CH_2)_{16}COOH$ or C18:0
Arachidic	Eicosanoic acid	$CH_3(CH_2)_{18}COOH$ or C20:0
Behenic	Docosanoic acid	$CH_3(CH_2)_{20}COOH$ or C22:0

Also, where chains are saturated, they are less reactive than chains of double bonds.

In a saturated fat, all the carbon atoms form a chain of single bonds, and all the binding sites are "fully occupied."

So, to recap, the main factors distinguishing different types of oil and fat are the chain length of the hydrocarbon chain and the degree of saturation.

So, we've seen how different types of oil can be formed from different fatty acids, different chain lengths, and different degrees of saturation. This variation in the composition of oils and fat leads to different chemical properties: For example, oil from soy beans has a lower melting point than, for example, rapeseed oil (canola oil).

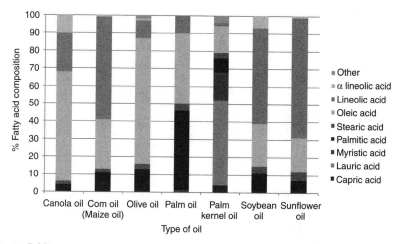

Figure 5.29
Fatty acid composition of different vegetable oil types.

Table 5.3 Percent by weight of total fatty acids of some common edible oils and fats

Oil or fat	Unsat./Sat. ratio	Saturated					Mono-unsaturated	Poly-unsaturated	
		Capric acid C10:0	Lauric acid C12:0	Myristic acid C14:0	Palmitic acid C16:0	Stearic acid C18:0	Oleic acid C18:1	linoleic acid (ω6) C18:2	Alpha linolenic acid (ω3) C18:3
Almond oil	9.7	–	–	–	7	2	69	17	–
Beef tallow	0.9	–	–	3	24	19	43	3	1
Butterfat (cow)	0.5	3	3	11	27	12	29	2	1
Butterfat (goat)	0.5	7	3	9	25	12	27	3	1
Butterfat (human)	1.0	2	5	8	25	8	35	9	1
Canola oil	15.7	–	–	–	4	2	62	22	10
Cocoa butter	0.6	–	–	–	25	38	32	3	–
Cod liver oil	2.9	–	–	8	17	–	22	5	–
Coconut oil	0.1	6	47	18	9	3	6	2	–
Corn oil (maize oil)	6.7	–	–	–	11	2	28	58	1
Cottonseed oil	2.8	–	–	1	22	3	19	54	1
Flaxseed oil	9.0	–	–	–	3	7	21	16	53
Grape seed oil	7.3	–	–	–	8	4	15	73	–
Lard (pork fat)	1.2	–	–	2	26	14	44	10	–
Olive oil	4.6	–	–	–	13	3	71	10	1
Palm oil	1.0	–	–	1	45	4	40	10	–
Palm olein	1.3	–	–	1	37	4	46	11	–
Palm kernel oil	0.2	4	48	16	8	3	15	2	–
Peanut oil	4.0	–	–	–	11	2	48	32	–
Safflower oil*	10.1	–	–	–	7	2	13	78	–
Sesame oil	6.6	–	–	–	9	4	41	45	–
Soybean oil	5.7	–	–	–	11	4	24	54	7
Sunflower oil*	7.3	–	–	–	7	5	19	68	1
Walnut oil	5.3	–	–	–	11	5	28	51	5

* Not high-oleic (mono-unsaturated) variety

Decomposing fats and oils

Where moisture is present, fats and oils can decompose. Heat also causes this to happen.

As fats break down, their triglyceride molecules form monoglycerides, diglycerides, and FFAs. When making biodiesel, we need to be aware of these components in our oil, and compensate for them accordingly.

One of the factors that makes certain types of oil desirable is a long shelf life; we have seen how saturated fats contain chains that are saturated with all their hydrogen bonds filled. It is possible to "saturate" a fat or oil artificially, through a process called hydrogenation, where hydrogen gas is injected into the fat or oil, saturating the double bonds, filling the binding sites of the carbon atoms, and producing an oil that is more durable, with a longer shelf life.

Monoglycerides

A monoglyceride is where one single fatty acid chain is left bonded to a glycerol molecule, the two others having become detached, FFAs, and their "attachment points" can be thought of as being "capped off" with a hydroxyl functional group (-OH).

Diglycerides

A diglyceride is much like the monoglyceride above, only two single chains, rather than one, remain bonded to a glycerol molecule.

> Mono and diglycerides and free fatty acids are all undesirable in our finished biodiesel; they are all symptomatic of an incomplete reaction. See Chapter 6 for more information on this.

Esters

We make esters from the reaction of an alcohol and an acid.

Esters consist of an acid, where an OH (hydroxyl) group has been replaced with an O (alkyl) group. We can see from Figures 5.30–5.34 how we start off with a simple ester like methyl methanoate, and can then grow the ester chain in length by adding units of hydrogen and carbon.

Compare the methanoates with methanoic acid, and look at what parts of the molecule are similar and different. Similarly, look at the ethanoate esters, and compare them with the ethanoic acid and again look at what parts are similar and different.

These esters are shorter-chain esters, however, the mono-esters that we talk about using as biodiesel have much longer chains.

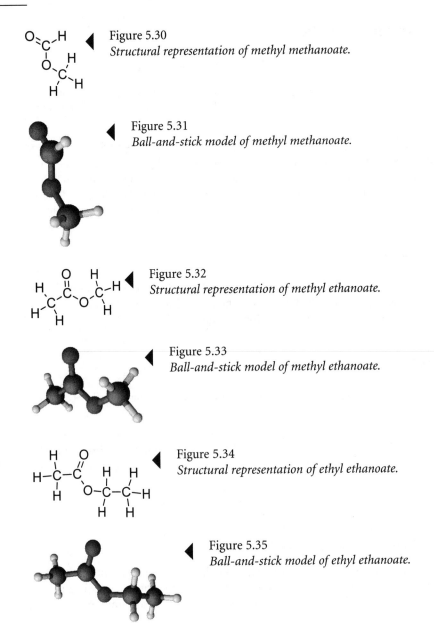

Figure 5.30
Structural representation of methyl methanoate.

Figure 5.31
Ball-and-stick model of methyl methanoate.

Figure 5.32
Structural representation of methyl ethanoate.

Figure 5.33
Ball-and-stick model of methyl ethanoate.

Figure 5.34
Structural representation of ethyl ethanoate.

Figure 5.35
Ball-and-stick model of ethyl ethanoate.

Esterification

Combining some of the concepts we looked at above we see that we can form esters from fatty acids and alcohols by a process called "Esterification." Esterification is the name given to

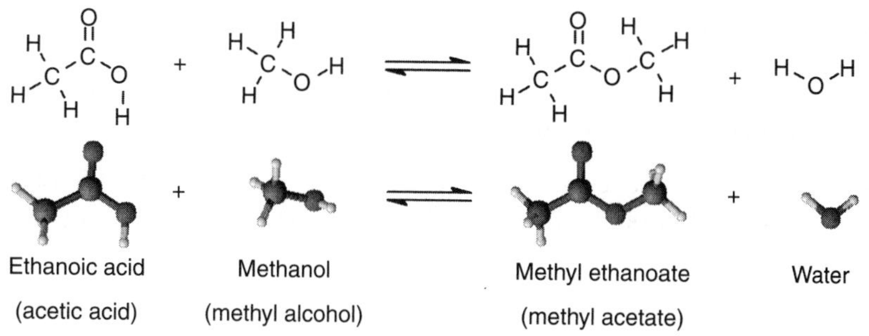

Ethanoic acid | Methanol | Methyl ethanoate | Water
(acetic acid) | (methyl alcohol) | (methyl acetate)

▲ Figure 5.36
The esterification process.

a chemical process that results in an ester being formed. This can be viewed in Figure 5.36 which shows a structural representation and ball-and-stick model of the esterification process. Put in very simple terms, when we combine an acid and an alcohol we produce an ester.

We are now going to begin to pull some of these concepts together. Now that we understand esterification and fatty acids, we can begin to look at our vegetable oils, which returning to the parlance of the chemist are called triglycerides.

In homebrew biodiesel production, we routinely use methanol as our alcohol, although it is possible to use other alcohols such as ethanol. Using methanol for the esterification reaction results in monoesters of methanol, which we can call monoalkyl methyl esters, or just methyl esters for short.

During the reaction, our "basic catalyst" (basic in this chemical usage means that it is a strong base, not using the word to mean simple!) breaks the chains in turn from the glycerol backbone. Once the chains have been broken off, the fatty acid chains react with the methanol in the methoxide mix to form esters.

The reaction is sensitive to temperature – we can apply a little heat to accelerate the reaction.

Table 5.4 Temperature sensitivity of esterification

Temperature (°C)	Temperature (°F)	Rate of reaction
21	70	4–8 hours
40	105	2–4 hours
60	140	1–2 hours

Be careful how much you try to "accelerate the reaction" – don't heat it any more than 140°F as methanol will "boil off" at 63°C (148°F), ruining your reaction and venting lots of toxic methanol fumes – best to keep it under 50°C (120°F).

Transesterification

Put simply, the transesterification reaction means taking one type of ester and turning it into another, for example, taking vegetable oil and turning it into biodiesel.

In this reaction we are exchanging the alkyl group of one kind of alcohol for another alkyl group from a different alcohol (remember, the glycerol backbone of the triglyceride [see below] is an alcohol!)

Triglycerides

Note that the hydrogens in Figure 5.38 have been omitted so that you can see the chains a little more clearly. In this model, we have instead represented the hydrogens by removing the "explicit" hydrogens – i.e., the bonds that we "know" to be there – instead, representing them

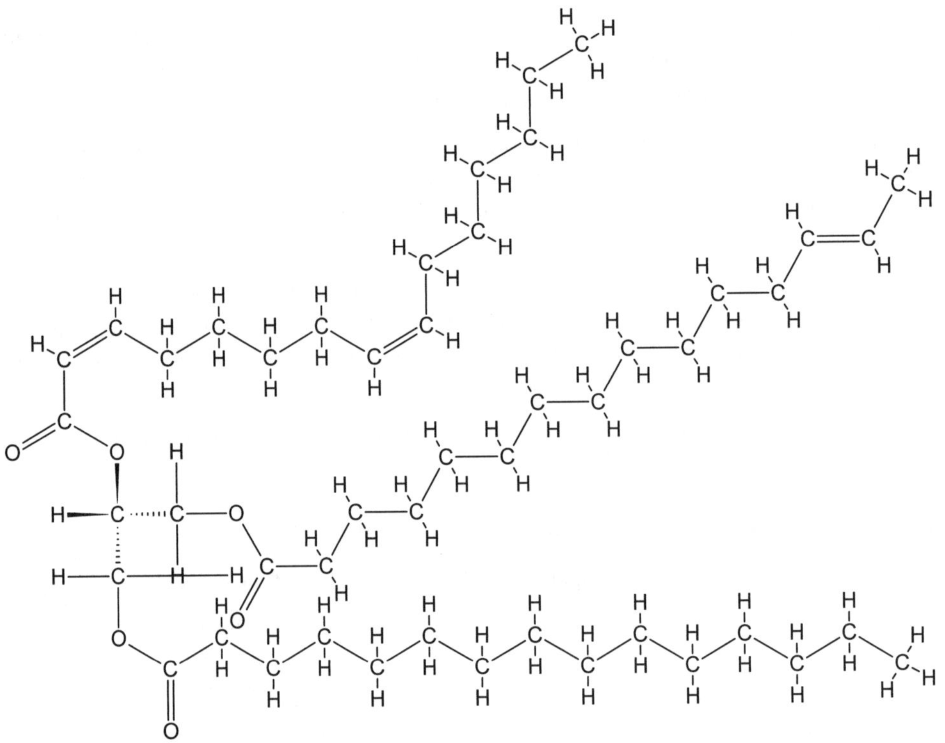

▲ Figure 5.37
Chemical structure of a triglyceride (expanded form).

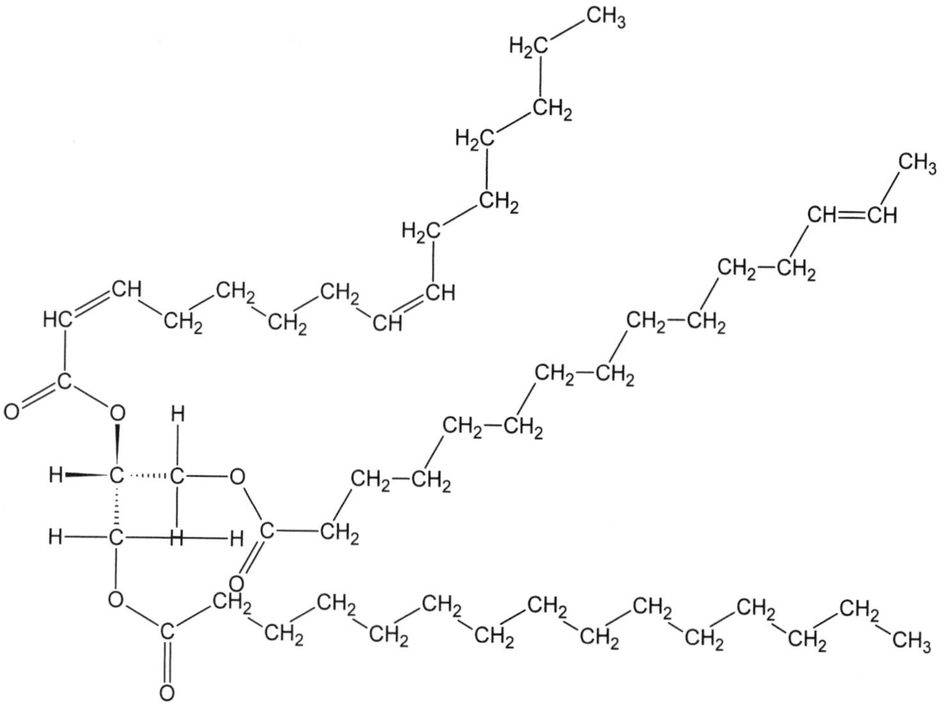

▲ Figure 5.38
Chemical structure of a triglyceride (condensed form).

with the condensed form of notation. If you feel more comfortable with being able to see all the bonds, see Figure 5.37.

One of the things that should be immediately apparent to you from looking at the figures is that the triglyceride molecules are much bigger than anything else we have seen in this chapter so far.

Because the triglyceride molecules are so large, and because of their complex shape, they form an oil which is very thick and viscous. When they are used as a fuel for the diesel engine this viscosity will inevitably cause problems.

One of the most common ways of making triglycerides more suitable for use as fuel for a diesel engine is to process them into biodiesel. This is most commonly done by reacting them, in the presence of a base-catalyst, with another alcohol, usually methanol, which produces methanol-mono-ester chains, our biodiesel. The glycerol backbone, Figure 5.40

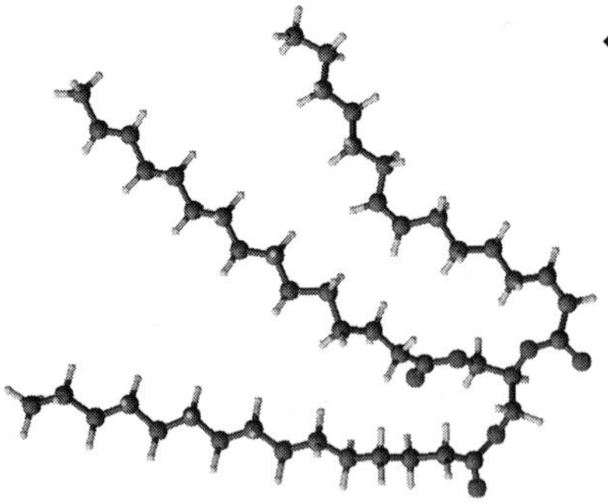

Figure 5.39
Ball-and-stick model of a triglyceride.

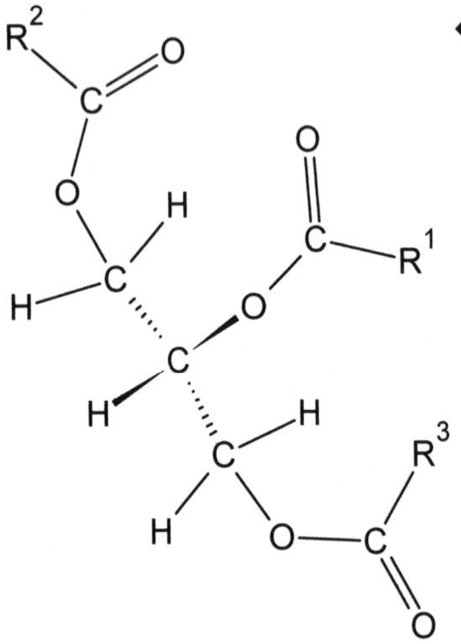

Figure 5.40
The "glycerol backbone."

is left unstable, so the base donates a hydroxyl group (—OH) to the glycerol backbone to produce glycerol (Figure 5.41).

Any free fatty acids present in our oil, and there will always be some, will also react with the base. This uses up some of our catalyst and results in soap and water, called sopification. As we shall see in Chapter 6 none of these are welcome when making biodiesel.

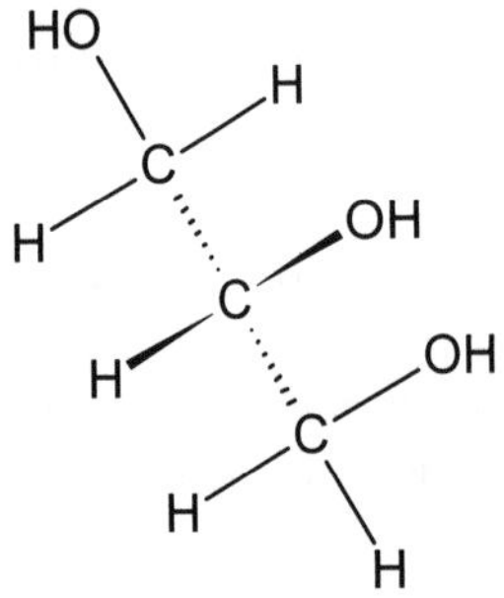

Figure 5.41
Structural representation of a glycerol molecule.

Figure 5.42
Ball-and-stick model of a glycerol molecule.

We've now looked at biodiesel chemistry from a theoretical perspective; hopefully, this chapter has provided a refresher on some basic chemistry, and helped you to understand some of the basic chemical building blocks and processes involved in the making of biodiesel.

In the next chapter we are going to take this knowledge and apply it in a practical setting, where things are never as straightforward as they are in theory, looking at the how-to of making biodiesel for real!

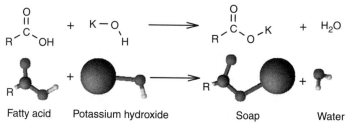

Fatty acid   Potassium hydroxide   Soap   Water

Figure 5.43
Sopification -the fatty acids react with the base to form soap and water.

Chapter 6

Fuel modification: making biodiesel

Introduction

Many people confuse running their car with vegetable oil with running their car on biodiesel. They talk about "converting it to run on biodiesel," when no conversion is required; biodiesel is a fuel suitable for any *unmodified* diesel engine.

A biofuel (sometimes called agrofuel or agrifuel) is a solid, liquid, or gas fuel made from biological material, usually plants. Fossil fuels are also also derived from materials that were once alive but which have undergone considerable change over millions of years. Generally speaking, biofuels are sustainable and fossil fuels are not. Examples of biofuels could be anything from wood to whale oil and include fuels such as vegetable oil, biodiesel, and some ethanol; examples of fossil fuels include gasoline, diesel, heating oil, bitumen, coal, and natural gas.

So, while vegetable oil is a biofuel, it is not biodiesel, which is a diesellike fuel made from vegetable oil and has undergone a chemical reaction, usually transesterification, to make it similar to mineral diesel.

Running your engine directly on straight vegetable oil (SVO) or waste vegetable oil (WVO) are dealt with in their own separate chapters, as is the chemistry involved in the making of biodiesel and the construction of biodiesel reactors or processors. Because the collection of the ingredient WVO is common to several techniques, it has its own chapter too. This chapter is primarily focused on the realities of making biodiesel.

As covered in detail in Chapter 5, making biodiesel involves replacing the vegetable oil's glycerol with another alcohol, usually methanol. Vegetable oil is made up of triglyceride molecules, each having a glycerol molecule with three fatty acid molecule chains hanging onto it; connecting them all together are ester bonds. If we think of glycerol as a viscous alcohol, and methanol as a less viscous alcohol, then what we are trying to achieve in the making of biodiesel is to replace the glycerol with methanol and thus make a less viscous oil. So, to the vegetable oil, the triglyceride, is added methanol and a catalyst. The catalyst breaks, or cracks, the ester bonds and the fatty acid chains become bonded to the methanol molecules, instead creating fatty acid methyl esters (FAME) or biodiesel, leaving behind the glycerol and the catalyst as by-products.

Unfortunately this is not the whole story. Unless you are using new oil, the feedstock will not only contain triglycerides, but also, among other things, free fatty acids and water. Heating the oil and prolonged use causes the triglyceride molecule to break apart. Some of the fatty acid chains become detached from the glycerol molecule and become free fatty acids (FFAs). Water also gets into the oil both during use and afterwards when it is stored for collection and disposal.

Both FFAs and water in your feedstock will complicate the reaction by making a soapy by-product. Whereas some soap is inevitable, a lot of soap is going to cause problems. Although more advanced methods of biodiesel production can cope with high FFAs, it is best to choose feedstock that is low in FFAs and water, especially when making biodiesel is a new experience.

Health and safety

Biodiesel production uses some dangerous chemicals, some of which can burn your skin, blind you if you get them in your eyes, intoxicate and poison you, or burn your house down. Don't let this scare you off, however; if you are careful, use common sense, and follow sensible procedures, you should not have any big problems.

You need to be very careful when handling the chemicals; use chemical-proof gloves, goggles, an apron, long-sleeved tops, shoes on your feet, good ventilation, and take care not to breath in fumes or dust. For detailed information, see Chapter 13.

Methanol is both very flammable and poisonous. It burns with an almost invisible flame, is almost odorless, and is intoxicating via inhalation and through the skin. Methanol can also burn your skin but you probably won't feel it burn you as it kills the nerve endings as it goes. Always work with methanol in a very well-ventilated area, never lean over containers of methanol, always wear chemical-proof gloves and eye protection, never have naked flames, and be very careful with other sources of heat, near methanol. And obviously, do I even need to say this; never smoke anywhere near methanol! Methanol vapor is heavier than air; it will collect on the floor and in hollows, and vapor protection masks are useless against it.

Lye, a biodiesel catalyst, is caustic and will cause nasty burns if you get it on your skin and could blind you if you get it in your eye. Always wear chemical-proof gloves, eye protection, and a dust mask when handling any catalyst and be careful.

The correct treatment for caustic burns is not to attempt to neutralize them but to wash with copious amounts of water and seek medical attention immediately.

Vegetable oil may seem rather innocuous when compared to methanol and lye, but it is dirty and containers of it are heavy. It will make floors and surfaces very slippery. Hot fat and oils are the biggest cause of domestic fires and, while we are unlikely to be heating much oil to that kind of temperature, it is worth being aware that vegetable oil is also a fire hazard.

While small furry creatures might be man's best friend most of the time, when it comes to making biodiesel, they are a definate no-no. No pets anywhere near you when you are working. Remember, while used cooking oil may seem repugnant to you, your dog and other animals, even wild ones, will probably find it very attractive; even when it becomes biodiesel, animals may still want to eat the potentially poisonous stuff.

All of the reactor designs in this chapter use electricity to power pumps or agitators. Ensure that if you are constructing these designs, and you are using grid power you are competent enough to understand what you are doing. If at any stage you feel unsure about wiring, consult a qualified, experienced individual to check your work. Water, other chemicals, and electricity aren't the best playmates in the world, so ensure that your wiring is "ingress protected" by using suitable fixtures and fittings that will not permit the ingress of liquids or dust.

Where to obtain the chemicals

Catalyst or lye

Sodium hydroxide (NaOH) was once relatively easy to get as drain cleaner and was very popular with biodiesel home brewers because of this. However, it is becoming hard to get in small quantities; one of the reasons for this is because it is used in the manufacture of methamphetamine, an illegal stimulant. Older biodiesel books recommend Red Devil drain cleaner but it no longer contains NaOH and, while it may well still be good at clearing your drain, it is no use at all for making biodiesel (or methamphetamine presumably).

If you can find NaOH, make sure you get powdered or granular, not the liquid type, and make sure it is 99% pure NaOH and that it is in a sealed container and has not been contaminated by air or water.

Potassium hydroxide (KOH) was always the preferable catalyst when making biodiesel, as it is both easier to use and more reliable. NaOH was only more common because of its easy availability. Now it is no longer easy to find, there is little reason to continue using it. KOH of purity 85% or more is usable for making biodiesel.

It is better to work with chemicals of a higher purity, as then you will need smaller amounts of reactant. Higher-purity chemicals cost a little more though. You really want to be working with chemicals of at least 85–90% purity or better.

It is possible to compensate for chemicals that are less pure by using a little more. There is a relatively simple method to allow you to judge how to compensate for chemicals of inferior purity; e.g., if you need 10 g of chemical and you are using 80% pure chemical, you will actually need 12.5 g.

To do that we divide 10 by 80 (this tells us what 1% is), and then multiply by 100.

$$10/80 = 0.125$$
$$0.125 \times 100 = 12.5 \text{ g}$$

Sources of small quantities of catalyst include fellow biodieselers, soap makers and soap maker suppliers, and leather tanners. Larger quantities can be bought from chemical suppliers and manufacturers; look up your local one on the Internet.

> The catalyst will react with the carbon dioxide in the air very quickly making it useless. It is also very hydroscopic; it absorbs water. Always keep your catalyst in an airtight container and don't buy too much in one go. Keeping it in ziploc bags is very helpful as you can squeeze almost all of the air out of the container even when it is half full.

Methanol is available in small quantities as antifreeze, model airplane fuel from model shops, or stove or barbeque fuel. However, not all of these products are necessarily methanol. Read the label to find out what it contains, and make sure it is at least 99% pure. Race car shops and drag race tracks can be helpful in the USA. Fellow biodieselers everywhere may be willing to help you. Larger quantities can be bought from chemical suppliers and manufacturers; again, look up your local one on the Internet.

Both small and large quantities of the chemicals required for making biodiesel have become available on eBay. While usally expensive, this can be a very simple method of obtaining the materials required, especially when you are just starting out; they can be ordered with a few clicks of a mouse and delivered to your door in the mail.

eBay is also a good place to look for other chemical supplies, such as tri-balance scales.

Making biodiesel

When making your very first biodiesel you could do a lot worse than simply buying a liter of new oil from the supermarket and using that. However, when making biodiesel in the real world you are almost certainly going to be using used oil (grease or WVO or UCO or what-ever you want to call it) as your feedstock, for both economic and ethical reasons. As mentioned above, making biodiesel from used oil involves getting to grips with its varying qualities. Repeated use and heating of the oil, sometimes the overuse and overheating of the oil, will have introduced FFAs and water and these and their by-products will be a problem unless you understand how to deal with them.

Titration

When you try to make biodiesel from used oil, the FFAs in it will react with your catalyst, making soap. This not only introduces soap into your biodiesel, which may be a problem during later processing stages, but also uses up the catalyst so it is not available to make biodiesel, resulting in an incomplete reaction.

To compensate for this extra use of our catalyst, we have to test the WVO for FFAs so we can add enough extra catalyst to neutralize them with enough left over for a good vegetable oil to biodiesel reaction.

Check out Chapter 4 for additional information on how to carry out titration.

Very heavily used oils will titrate at higher values and require a greater amount of catalyst to neutralize the FFAs. There comes a point where you will no longer easily yield decent biodiesel, because of the amount of soap you will make, and you may get some very dis-heartening failures and emulsions.

How high is too high a titration depends on which catalyst you are using, NaOH will start to be unproductive at about 5, whereas if you are using KOH you may be able to go as high as 8. However, it is better to leave oil titrating higher than 4 alone and not to collect it in the first place, at least not until you are a lot more experienced biodiesel maker and can consider using the acid/base method (see at the end of this chapter).

Water in your WVO

Used cooking oil will contain some water and you want to remove it before you convert the oil into biodiesel. The easiest way of dealing with this is by collecting only relatively dry WVO in the first place (see Chapter 4); but in the real world you will need to dewater the oil after it has been collected.

Testing for water is a simple procedure, if rather crude. Take a small sample of your oil in a pan or a ladle and, well away from any flammable substance such as methanol, heat it up. When the oil approaches the boiling point of water, it will spit and writhe if there is any

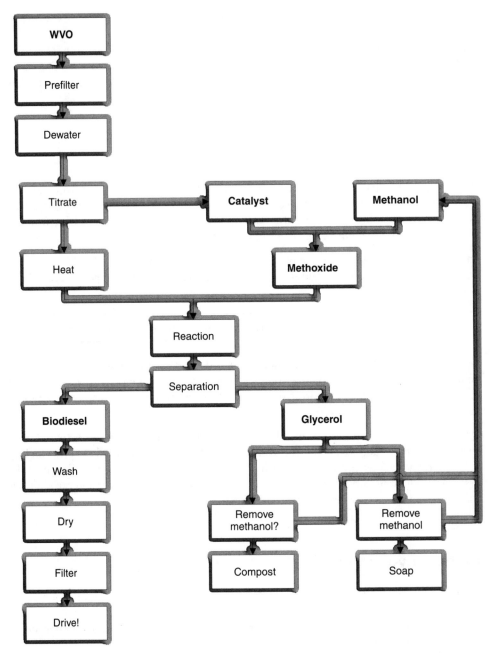

▲ Figure 6.1
Flowchart, from vegetable oil to biodiesel (simplified).

water present. As to how wet the oil is up to you to decide. It is a matter of experience to gauge, but whether you are planning on making biodiesel or using the WVO in your converted engine you will want to remove just about all of this water.

Chapter 9 has a section on dewatering oil. Two procedures are described: One involves heating up all the oil and boiling off the water; the other is to encourage the heavier water to sink under the dryer oil. Follow these procedures if there is water in your oil, and then retest for water.

Test batches

Making a test batch is something you are going to be doing each time you get a fresh batch of oil, as it will help you avoid all kinds of large-scale disasters. Even as a beginner, with no large-scale reactor, this is something you should be doing regularly: deliberately making different mixtures of waste oil and new oil (to change the FFA ratio) and converting it to biodiesel. There is no substitute here for experience; making mistakes, even deliberate ones, and overcoming problems is time well spent and will pay off in the long run.

All the safety procedures and equipment apply here: gloves, goggles, long sleeves, good ventilation, and good practice are a minimum.

Requirements

- Some used vegetable oil that has been dewatered.
- Some catalyst.
- Some methanol.
- Several glass jam jars or similar, clean and dry, with good fitting lids.
- Several plastic cola bottles – the 2-liter ones, clean and dry, with good fitting lids.
- A funnel to fit top of cola bottle.
- A small measuring jug.
- A teaspoon.
- A thermometer; jam or candy making ones are ideal, able to read from around 40–100°C.
- Saucepan, plenty big enough to fit the oil in.
- Some way of heating the pan, preferably an electric hot ring rather than gas.
- A scale, accurate to 0.1 g. Digital ones can be suitable, a tri-balance is good.
- A titration kit (see titration above).
- Some way of labeling the containers. A biro and adhesive labels or a magic marker will do. You will need to remember what the oil titrated at and what you put into the jars; methanol looks just like methanol with 5 g of catalyst in it, which looks just like methanol with 7 g of catalyst in it!

You will also need:

- Some chemical-resistant gloves
- Some goggles
- A well-ventilated space
- Shoes on your feet
- An apron
- Long sleeves

Remember: Keep children and pets out of harm's way.

Method

- Take a 1-liter sample of the oil.
- Put on your protective gear.
- Perform a titration test on the oil. Do it several times to ensure you got a good result.
- Gently heat the oil to around 45°C (130°F); check, using a thermometer. It won't take long; be careful not to overheat it or you may make more FFAs and you don't want to be adding methanol to oil over around 50°C as it will begin to evaporate.
- Measure about 220 ml of methanol into a jar, being careful not to breathe in any fumes; 220 ml is 22% of a liter – recipes vary for biodiesel at between about 20% and 25%. Only about 15% is used in the reaction; the rest drives the reaction forward and helps ensure a good conversion rate.
- Now weigh out your catalyst. Close its container as soon as possible; remember the catalyst will react with the air, making it useless. Use 5 g per liter of oil for NaOH and 7 g per liter of oil when using KOH for unused oil, plus the titration number. So, if the oil tritrated at 2 and you were using KOH, that would be 9 g of catalyst per liter of oil (7 g + 2 g). If it titrated at 3 and you were using NaOH, then that would be 8 g of catalyst per liter of oil (5 g + 3 g).
- Put the catalyst into the methanol immediately, being careful not to breathe in any of the fumes, and put the lid on the container, labeling it with the amount of catalyst and methanol as you go. Swirl the mixture *gently* in the jar until the catalyst is dissolved. You may notice the mixture getting warm at this point; this is because an exothermic reaction is taking place between the two. This is now "methoxide."
- Check that your oil's temperature is not above 50°C (140°F) and pour 1 liter of it into the cola bottle, then add your "methoxide," again being careful not to breath in any fumes.
- Put the cap on securely and shake the mixture vigorously for ≥10 minutes.
- Leave the mixture to stand for several hours.

If all went well

If you see clear separation, you have probably made some biodiesel, but only further testing will show how good it is. The bottom layer should be roughly 20% of the total and dark in color; this is generally referred to as "glycerol" but is more accurately called the by-product as it is only about half glycerol, with the rest being most of the reaction's leftover methanol and soap.

If you appear to have successfully made biodiesel, then well done! Let's proceed to the quality testing stage and find out what you really made.

If all went not so well

If you don't see the glycerol layer, or it is too small, then something has gone wrong some-where. This is fine too; this is why we are doing a test batch and you will learn from your mistakes. Go back over what you did and look for errors. If it did not separate or not enough

▲   Figure 6.2
Separation: a cloudy light-colored layer sits on top of a much darker layer. If you look carefully, in this photograph you can also see an intermediate soapy layer.

glycerol settled out, then you have an incomplete reaction. Maybe you used too little catalyst: check you titration and try again, or indeed reprocess the test batch.

If a third layer separates out, then you have made a lot of soap; most likely you used too much catalyst or did not remove sufficient water or just used oil which was too full of FFAs. If you appear to have made gloop, then you definitely made a lot of soap and the batch has become an emulsion. This is again too much catalyst or water, or both. Aren't you glad you only made a test batch and did not jump straight into reacting all 100 liters!

Do not be disheartened. Do not slip into a state of morose melancholy. There's always tomorrow, and if there's one thing we all learned from high-school chemistry, experiments rarely go the way we want them the first time. Re-read this chapter and the previous chapter on biodiesel chemistry and see if you can analyze what went wrong.

Ready to scale it up?

The small amount of biodiesel you've just made isn't going to take you on a particularly mean road trip; in fact, you'll be lucky if you manage a few blocks, let alone into the next state or all the way across the country … unless you live somewhere really small like Liechtenstein.

If it passes the quality testing below and you have still not made a scaled-up batch, time spent here, going over the process several times with different samples of oil, is time very well spent.

If you have run a few test batches, maybe it is time to try to scale things up a bit and see if you can repeat the process on a larger scale. Read the scaling it up section of this chapter and the building a biodiesel processor chapters. Remember, keep making test batches as you go, and take samples at each stage and test them before you run the next stage on the whole batch. You may regret it if you don't and end up with 100 liters of gloop.

Is this not biodiesel then?

In years gone by, biodiesel homebrewers were happy at this point; some people go all misty eyed when reminiscing about times gone by when standards were lower and folks were more easily pleased – something had separated out so they had made "biodiesel." As for its quality or contamination, they were not all that concerned. Truth is, your biodiesel's reaction may well have only partially completed; it could still contain a large amount of triglycerides, diglycerides, and monoglycerides; and is almost certainly contaminated by significant amounts of methanol and soap. We want to test for good conversion and, if satisfied, clean the product before using it. Bad biodiesel gives the technology a bad name; in the long run, it's not going to do your vehicle a world of good and it's only going to destroy your faith in what is a good system – spend the time to get it right, and your wallet will benefit from reduced repair bills.

Quality testing

Separate the glycerol layer from the biodiesel layer in your test batch bottle by gently pouring off the biodiesel into a new clean container, leaving the glycerol at the bottom of the old container. Think about pouring a layer of fat off the top of a jug of gravy that has been left to settle: Do it nice and slow, and don't let the liquids mix. If you are running a big batch, not a test batch, then leave it to separate and settle for a good 12 hours before draining the glycerol by-product, and then continue.

Visual inspection

Give it the eyeball. The first inspection of your fuel is a visual one: You have probably already done this, and unfortunately it does not yield an awful lot of information. Biodiesel, good and bad, has a fairly wide spectrum of colors. When we are talking about cloudiness, we are not talking about its color, we are talking about your ability to see through it. Unwashed cloudy fuel can indicate that you have soap in there; some soap is to be expected and can be washed out. A third intermediate layer between the biodiesel and the glycerol is probably an indication you have made more soap than was necessary, possibly by adding too much catalyst to the reaction. Warm biodiesel that appeared to be clear can become cloudy when it cools down. Nice clear biodiesel could be an indication you have made quality fuel, but could also be an indication you had low conversion; you need to test it further.

Wash test

What you will need: access to clean water; a clean bottle or jar with a lid that fits; some litmus paper or a pH meter.

You can predict a problem in the next stage: washing the biodiesel with a quick wash test. Place a sample of your biodiesel in a clean jar and add a roughly equal amount of water. Put on the lid and shake vigorously until the two form an emulsion, and then leave them to separate. They should separate into two distinct layers within about 10 minutes, biodiesel on top and cloudy water underneath. However, if your reaction was incomplete (and you have a lot of diglycerides and monoglycerides, which are emulsifiers) or you have a lot of soap (which is also an emulsifier), the emulsion will hold for longer, maybe what seems to be permanently.

Washing will also remove any leftover methanol and catalyst from your reaction.

There is a myth that well-made, well-converted biodiesel will not form an emulsion and that if you have an emulsion you have bad fuel. All biodiesel made from waste oil in a base/base reaction will contain some soap and soap is an emulsifier. Enough soap in your biodiesel, well converted or not, will potentially produce an emulsion. However, if you have used the acid/base method (see at end of this chapter), you will not have any soap; if you have good conversion using this method, you only have biodiesel, which should not form an emulsion with water on its own.

▲ Figure 6.3
  Washed biodiesel. After several washes, clear, sparkling biodiesel sits on top of clear water.

Just soap or an incomplete reaction?

To see if it is an incomplete reaction or just soap that is causing your emulsion, first carry out a wash on the sample several times (gently initially), each time waiting for separation, removing the water, and washing again.

When the wash water remains clear after the wash, and its pH is neutral (pH = 7), you have successfully washed out all the soap (along with the other water-soluble impurities). If you wash it again and the emulsion holds for significantly longer than a few minutes, you probably have an incomplete reaction, but we can test further for this.

Testing for diglycerides and monoglycerides in your biodiesel

Between being a triglyceride, with three fatty acid "arms," and being a biodiesel molecule, there are intermediate stages for the oil molecule: A "two-armed triglyceride" is a diglyceride, and a "one-armed diglyceride," is a monoglyceride (or more correctly, monoglycerine).

Commonly called the 3/27 test, the following procedure tests for the presence of diglycerides and monoglycerides in your biodiesel; or to put it another way, this is a good test to check how well converted your fuel is.

Measuring carefully, dissolve three parts biodiesel in 27 parts methanol in a clean jar and give it a good shake. If no oil drops out of the solution, you probably have well-converted fuel. If some oil drops out, these are triglycerides, which are not soluble in methanol, and it indicates the likely presence of diglycerides and monoglycerides too.

If you have poor conversion and this is a large batch, you may want to reprocess it; if it is a test batch, you will want to run the test again.

Testing the specific gravity of the fuel

A liquid's specific gravity (SG) is its density, or how much 1 liter of it weighs: e.g., 1 liter of pure water weighs exactly 1 kilogram; in fact that is its definition.

Biodiesel should have a density of about 887 g per liter, about 11% less dense than water, and you can test for this in two ways: Either use a winemaker's hydrometer or very carefully measure out liter of biodiesel and accurately weigh it.

Unfortunately, this does not tell you much about the quality of the fuel; unwashed biodiesel contains any number of impurities that will throw off your measurements, and washed, but poorly converted, biodiesel may well have an SG of around the right amount for well-converted biodiesel despite it containing tri-, di-, and monoglycerides.

"Proper" quality testing

Ultimately, the only real way of testing the quality of your fuel is to get a laboratory to do it for you, which is expensive and generally unnecessary for the home brewer. The techniques described in this book ought to be sufficient. Tests cost from US $100 upwards using machines such as a gas chromatograph. Other lab tests also exist but are usually even more expensive. If you are interested in getting your fuel "properly" tested, you could do a lot worse that asking chemist friends and university students for their help and see where this takes you.

Making usable batches of biodiesel – scaling up

There is a bit of a jump from making test batches in a bottle to making large batches in your own processor, but everything you have learned so far still applies and you will still want to make test batches as you go to check your titration and your recipe. (See Chapter 7 for more information on making processors and reactors.)

Once you have checked your recipe by making a test batch, scale up your ingredients proportionally. Read the instructions for your reactor and proceed. Your reaction should be allowed to run for at least 1 hour for oil at 50°C (120°F) to ensure good conversion, longer if the temperature is lower or your reactor is poorly insulated.

After you have finished the reaction, check and test for conversion; don't assume that because it worked in the test batch that it would be successful here.

Washing your biodiesel

Many old biodiesel books will tell you that the washing stage in biodiesel production is unnecessary – they are very wrong. Washing out the impurities in your biodiesel is important; the leftover catalyst is corrosive and will damage your engine. Methanol is a good fuel for drag-race cars but not your diesel engine. Washing out these two will also stop any further reaction taking place that would otherwise continue creating small amounts of glycerol in your fuel and block filters and damage engines. Washing also removes any soap in your biodiesel that did not settle out with the glycerol layer and that may also damage your engine if not removed.

Always perform a wash test on your fuel before washing your entire batch to test how it will perform in the wash, and then a 3/27 conversion test. If your fuel appears to be poorly converted, reprocess it before proceeding to washing.

Initial washing is usually a gentle mist or bubble wash in order to remove much of the soap but with a lowered risk of making an emulsion.

Bubble washing occurs when a few gallons of water are added to the biodiesel. The water will sink to the bottom of the container and the biodiesel will sit above it. An aquarium air-stone (check the one you have chosen does not disintegrate in the biodiesel) and pump is used to bubble air through the water and the biodiesel. Water is carried with the bubble through biodiesel, dissolving impurities as it goes. When the bubble reaches the surface, it pops and the water sinks back through the biodiesel, collecting more impurities as it goes back down.

Mist washing involves spraying a fine mist of water over the surface of the biodiesel. The tiny water droplets sink through the fuel and dissolve impurities as they drop down. The water is then either drained off or recirculated.

The advantages of bubble washing and mist washing are reduced complexity. They are easy to make, and you reduce the risk of making an emulsion from your whole batch, especially on your first wash when the soap concentration is highest.

The main disadvantage is the time it takes to complete; you may need to run the bubble washer for 8 hours or more (overnight is common), and you will need to change the water and run it again several times until your wash water comes out clear and pH neutral.

With the mist washer you don't generally change the water; it only does one pass and is allowed to drain away. Although this is quicker than the bubble washer, where you have to stop the process and change the water, it is also likely to use a lot more water.

Other washing methods exist for washing biodiesel. They are generally more aggressive, such as vigorously stirring the water-biodiesel mix. These methods are much quicker than the gentle approaches, which can take several days to complete, but unless you are very confident of low or no soap content (i.e., you have used new oil, or your used oil titrated very low, or you used the acid/base method described in this chapter), you run the very real risk or emulsifying your whole batch.

In addition, the equipment is generally more complex and expensive and therefore not so appealing to some homebrewers.

There is probably room for using a combination: gentle washing methods first and more aggressive methods when you are confident that the chance of emulsification is sufficiently low. This will speed up overall washing times and reduce water consumption.

How to break an emulsion

This should not happen, if you have been taking samples and testing as you go, but occasionally it happens anyway. Maybe your test results were unreliable, or you were just unlucky this time. Whatever, don't panic!

The easiest solution to emulsified fuel/water is time; given enough time, an emulsion will often break of its own accord and settle into two distinct layers.

As you may recall from Chapter 5, glycerol is an alcohol. Alcohols can be used to break emulsions. If you have an emulsion and you have given it time to break of its on accord, you can use glycerol to break it.

First, ensure you have drained off all the water you can. Then add glycerol and, watching out for methanol fumes, stir it in. If the emulsion does not break, mix in some more until it does. Once it has broken, drain off the glycerol and retest. If you have good conversion, wash again, only more gently this time!

Other ways of breaking an emulsion include simply continuing to mist wash the emulsion until it breaks; this usually works but uses an enormous amount of water.

Some people add acids or salt to cause separation. These methods also work but acid in your fuel will damage your engine and cause your fuel to be contaminated by FFAs. Salt can be hard to wash out and is also not good for your engine.

Some people use heat to break the emulsion. This again works but needs to be done with caution because the water in, or underneath, the emulsion may boil, causing dangerous pockets of steam. Also, the mixture will still contain methanol, which will vaporize and poison you; be careful!

Drying your biodiesel

After washing your biodiesel, it will most likely have water suspended in it and will make your previously clear but unwashed biodiesel appear disappointingly cloudy. Suspended water is different from having water dissolved in the fuel; suspended water is relatively easy to remove and will damage your engine if you skip this stage.

Leaving the fuel to stand in an open container is the simple method of removing the suspended water: Given time, it will clear of its own accord. This process can be sped up by pumping the fuel out from the bottom of the container and back in at the top, sometimes by squirting it down and around the side of the container, while actively ventilating and circulating air across the top of it with a fan.

The biodiesel is generally considered dry enough if you can hold a jar of it up to the light and easily read newspaper print through it.

Filtering the biodiesel

After making your biodiesel, testing its conversion ratio, and washing and drying it, one final stage of processing is needed before it is ready for use in your engine: It needs to be filtered. Even if you filtered it before it was converted from oil to biodiesel, it is a very good idea to filter it again to remove any particles of dirt it may have picked up from or during its processing. Most engines come with a 10-micron filter, some lower, filtering your fuel to 10-microns (μm) before it goes into your fuel tank is essential.

Recovering methanol

In making your biodiesel, you added up to 250 ml of methanol for each liter of oil. Only about 150 ml of this was used up in the reaction. The rest was there to ensure the reaction went in the right direction, from oil to biodiesel, and to keep conversion rates high.

By leaving the glycerol to settle in the reactor for 12 hours or more, most of this excess methanol will be in the glycerol layer, along with glycerol, catalyst, and soap. Recovering the methanol from the biodiesel is generally not considered economical, since there is not much there to recover; instead, what is there is washed out. However, recovering the methanol from the glycerol is often considered worthwhile.

The process is to heat the glycerol up to around its boiling point of 65°C(149°F). The boiling point of methanol-in-solution is not the same as pure methanol, however, and we don't want to contaminate what we collect with water if we can avoid it, so methanol recovery stills are usually run at around 55°C(131°F).

See Chapter 7 for how to make a methanol recovery still.

How to achieve high-quality biodiesel

Here is a quick checklist for making good biodiesel:

- Choose oil with low FFA content.
- Ensure the water content in your oil is very low, close to zero; remember, oil + catalyst + water = soap.
- Use the correct amount of catalyst: not too much and not too little.
- Use sufficient methanol.
- Ensure your oil is hot enough, and is mixed vigorously enough for long enough. Oil at 50°C should be mixed vigorously for at least 1 hour, preferably longer. For each drop of 10°C you can add another 1 hour to this; 2 hours for 40°C, 3 hours for 30°C. If in doubt, mix for longer: You can't overmix, but you can underdo it.
- Don't lose too much methanol during mixing. Reaction temperature should not go above about 55°C as the methanol will begin to evaporate; it boils at around 65°C; don't overheat your reaction and think about methanol recovery.
- Check for good conversion.
- Wash the fuel thoroughly.
- Dry the fuel well.
- Filter to at least 10 μm.

Other methods of making biodiesel

Single-stage transesterification is not the only way to make biodiesel. There are a number of methods of converting used oil to biodiesel, some breaking the reaction into multiple stages and others using straight esterification. Other ideas are way beyond the scope of the homebrewer, such as catalyst-free supercritical methods.

Two-stage reactions

Two-stage base/base

As discussed earlier, the relatively large quantities of methanol present in the reaction compared to the amount used in the reaction (25% compared to 15% typically) is there to drive the reaction in the right direction, from oil to biodiesel (the reaction is entirely reversible) and to keep conversion rates high.

The relative proportions of methanol and glycerol to each other (they are both alcohols remember) is what controls the direction of the reaction. When the glycerol and methanol approach equilibrium, the reaction will slow and eventually stop. So long as there is a lot more methanol than glycerol, the reaction goes in the right direction, and conversion rates are high.

In a two-stage reaction, some of the glycerol is removed part way through the reaction; this helps to push the reaction in the right direction, and generally conversion rates are higher. It also adds the possibility of reducing the amount of methanol used – some people claim to have achieved good conversion rates with significantly less than 20% methanol using this method.

The method is to split the methoxide into two equal parts and initially add only one-half to the reaction. The process is carried out as normal, with the reaction run for 1 hour and the glycerol allowed to settle out. The by-product is removed; the second half of the methoxide is added to the reaction; and the reaction continued as normal.

This method produces better conversion most of the time and uses less methanol use.

Two-stage acid/base

In an acid/base two-stage process, the FFAs in the used cooking oil are converted into biodiesel as well as the triglycerides.

An acid/base is used in the first stage to convert the FFAs to biodiesel. In the second stage, the rest of the oil (the triglycerides) are converted to biodiesel as well using a lye catalyst.

Because the FFs have been used up in the first stage, there are none present in the second stage to react with the catalyst, so there is no soap.

This is a superior reaction, creating more biodiesel and no soap. It is often very misleadingly called the "foolproof" process. While not out of the reach of most homebrewers, it is a complex procedure, as you are handling concentrated acid and a superior reactor needs to be built.

Buying biodiesel

You don't have to make it of course, you can always just let someone else do the hard work and you just buy it! It is available from your local biodiesel cooperative and even from garage forecourts in some places.

Biodiesel blends

B5, B20, B100 refers to the biodiesel content of the fuel as a percentage: B5 is 5% biodiesel and 95% mineral diesel; B20 is 20% biodiesel; B90 is, you've guessed it, 90% biodiesel; and, finally, B100 is not a blend at all, it is 100% biodiesel with no mineral diesel added.

▲ Figure 6.4
Buying biodiesel at the Solar Living Institute in Hopland, California.

Many vehicle manufactures permit a B5 blend so long as the biodiesel meets the necessary standards; some manufactures allow a higher blend, and some even allow B100. It is very common for homebrewers to add mineral diesel to their biodiesel in the winter to lower its cloud point, the point at which it begins to freeze, to make a more winter-hardy fuel.

Biodiesel available in regular fuel station's forecourts is often called green diesel or similar, and is almost always a blend of mineral diesel and biodiesel: usually B5, a 5% biodiesel to 95% mineral blend. In addition, this forecourt biodiesel is usually made from virgin oil.

In early 2008 a biofuels trading scam was uncovered. Perfectly legally, companies were importing biodiesel into the United States from Europe, a small amount of local biodiesel was blended, and then the cargo was exported from the United States back to Europe. This enabled a significant subsidy be claimed from the U.S. government, so the fuel was then salable at below European prices; up to 10% of U.S. exports at the time were estimated to be part of the scam. It was uncovered as part of an investigation into why, with biofuels fetching high prices in Europe, European biofuel makers were struggling to make a profit.

B5 blends have been the norm in France for many years, included in their fuel as a lubrication replacement for the sulfur removed to make low-sulfur diesel. In the UK, the renewable fuels obligation means that a percentage of all fuel sold at the pump must be composed of biofuels.

Buying biodiesel from the smaller producer

Buying biodiesel from the smaller guy is an excellent thing to be doing; keeping your money local, supporting the local entrepreneur, minimizing transport miles, and doing the right thing while not having to deal with any of the mess or hassle of making your own on converting your engine to SVO.

However, it pays to be a little cautious. Armed with the information in this book, you should be able to make informed choices, ask the right questions, maybe take a sample home with you for testing (these tests work well on other people's fuel as well as your own!) and don't take anecdotal evidence, "my engine runs fine on it," as good enough.

Figure 6.5
Private biodiesel storage tank in California.

When biodiesel is not biodiesel at all

In the first few years of the 21st century quite a number of small biodiesel outlets appeared across the UK, claiming to be selling "biodiesel" to the public. When pressed as to what exactly it was they were selling, it turned out to not be FAME biodiesel at all but their own "secret recipe"; this subsequently turned out to be a mix of paint thinners, mineral diesel, and WVO. While the authors have no problem with you running your own vehicle on whatever you like, these people were deliberately misleading well-meaning members of the public into buying what they assumed was FAME biodiesel when it was, at best, an experimental fuel and almost certainly not what the customer thought it was; this may have done untold damage to thousands of innocent people's engines, who knows.

For the last word on homebrew biodiesel, we're big fans of Maria "Girl Mark" Alovert's homebrew book – the "Biodiesel Homebrew Guide." Although self-published, this book is a thorough guide to biodiesel production and is available online from www.localb100.com/book.html.

Chapter 7

Biodiesel reactors and processors

Once you have become familiar with making biodiesel in small batches and you have found a good source of used oil, methanol, and catalyst you can think about scaling things up. This is not to say you no longer need to make test batches; you should always run tests as you go and double check before proceeding.

Biodiesel processors are as unique as the people who make them; even bought off-the-shelf reactors soon become modified and personalized. Your design will depend on your budget, what country you live in, what parts are available to you, what you find in the scrap yard, how much space you have, how much oil you can get hold of, how much biodiesel you are expecting to make, and how handy you are.

In the simplest terms, your processor and processing regimen will need to be designed to achieve the following:

- Remove the larger bits from your used oil.
- Dewater the used oil.
- Heat the oil to around 50°C (122°F) and hold it there. Usually an electrical heating element and insulation are needed.
- Fully dissolve the catalyst in the methanol, making methoxide, safely.
- Slowly and evenly add the methoxide to the heated used oil. Adding it too fast may cause it to be not distributed throughout the oil and, in the worst case, form a separate layer on top of the oil.
- Keep the methoxide evenly dispersed in the oil while the reaction is taking place.
- Mix and agitate the used oil while the reaction is taking place.
- Separate the glycerine and other by-products from the biodiesel after reaction has completed; use a settling tank, often the same tank as for the reaction.
- Remove the methanol from the biodiesel, usually water washing in a wash tank, but sometimes by distillation.
- Removing the soap and catalyst from biodiesel: water washing in wash tank.
- Dry the biodiesel to remove its suspended water: usually done in the wash tank.
- Filter the biodiesel: down to 10 μm or better.

Where to put the processor?

As you know, making biodiesel uses some dangerous chemicals, poisons, and flammable materials. Add to this heat and electricity, and you will see why you will probably want to choose to put your reactor in a building not attached to your house, such as an outbuilding, shed, or barn.

The risk of fire is real with hot oil; in fact, chip pans are the number one cause of domestic fires. In addition, "empty" tanks with methanol vapor in them are potentially explosive. Methanol vapors are toxic. So long as you are careful, the chances of your reactor bursting into flame is not likely, but the chances of burning your house down or poisoning your family are still very real if you process biodiesel in your house. You will also probably be making mistakes, maybe some funky smells, and possibly even a bit of a mess too.

The ideal space will need to be away from the house, reasonably sheltered from the elements; it will need a mains electricity supply, a water supply, and lighting, and be well ventilated.

Electricity

Unless you are very creative, you will need electricity to heat the oil and to pump it about. The size of your processor will determine how much amperage you need to allow for.

> Electricity and fluids don't make for an awesome party – unless your middle name is "Danger." You're going to be sloshing fluids around, and the chances are, if your wiring is not sealed, liquids are going to get in there. The first point is to make sure that all your wiring is tip-top. Use junction boxes and switches with the correct "Ingress protection" rating for dust and liquids. This will stop the chemicals you are using to make your biodiesel from penetrating your wiring. Ensure that any work you do is checked over by a competent person before you fire it up, and ensure that you check any local codes on what electrical work you can legally carry out in your country/state/region.
>
> A suitable earth-leak breaker MUST be installed to protect you from electrocution.

Electrical water heaters consume several thousand watts: In Europe this will be more than 10 A at 230 V; in the USA this will be more like 20 A if you are using a 110 V supply. This is before you have included the processor's pump or the other electrical demands of your processing shed such as lighting. You will need to ensure your shed's electrical connection is up to the job.

Water

Water supply will usually come from your domestic connection, from your local water supply company. This is fine, but do think about the cost, in both money and energy.

Water supply utility companies go to a great deal of effort and expense, in both money and energy, cleaning dirty water to the point where it is consistently 100% safe to drink. Consider using less of it when making biodiesel or maybe using collected rainwater.

Lighting

Unless you plan to make biodiesel in the summer months or the daytime only, chances you will need lighting in your new biodiesel lab. Lots of lighting is good to help avoid accidents. Refer to the above section on electrical work – ensure that your wiring is up to the job, and the light fittings you are using are splash-proof, for when you have chemicals sloshing about.

Ventilation

Very good ventilation is absolutely essential; methanol is very toxic, very flammable, and potentially explosive. Just having a window open is not good enough; you need active ventilation, fans to extract fumes away and out of your space. Remember, methanol vapor is almost odorless, so you won't smell it before it is too late, and it is heavy, so it will tend to sink to the floor and collect in any hollows.

Bunding and spills

Managing and containing possible spills is always a good idea, and possibly essential in order to meet local legislation, and yet is often overlooked. Bunding is a container large enough to catch the largest possible spill. Take the biggest container you have and imagine if it leaked overnight or, worse, fell over. What, if anything, would contain the spill? Where might it end up? Polluting a watercourse or a storm drain with something as innocuous as vegetable oil is a serious environmental matter which could result in a hefty bill or even you being prosecuted.

Designs of spill basins vary greatly, from homemade buckets and drip trays under individual containers to purposely made interlocking floor tiles or low walls built around a concrete floor to turn the whole workshop into one giant bund.

Reactor materials

The best material for building your reactor has got to be stainless steel; unfortunately, it is very expensive, becomes "work hardened," and requires a specialist to weld it.

If you are going to use the acid/base method, stainless steel or plastics are just about your only choices, as other materials will react and corrode rapidly when exposed to the acid.

Steel tanks and plumbing parts, "black iron" gas pipes, are plentiful, cheap, and relatively easy to work with and can be welded by a semiskilled person.

Copper is also an option, although it can sometimes be expensive (the price of copper fluctuates a lot). Copper plumbing parts and copper water tanks are very common, it can be soldered and brazed by a semiskilled person, and is easy to work with. Some people worry that copper and solder may interfere with your biodiesel reaction or reduce its shelf life; nevertheless, it is a very common material because of its easy availability.

Plastic components are available; one can even buy ready-made HDPE (high-density polyethylene) conical bottom tanks that seem ideal for making biodiesel processors because they make good vessels for separating fluids cleanly; e.g., your glycerol from your biodiesel. Plastic is also relatively inexpensive, easy to work with, and often semitransparent; however, it is almost always best avoided for making biodiesel processors.

> Plastic components are not very tolerant of heat; at best they become soft and at worst they can melt entirely. They regularly suffer from annoying leaks (because you cannot get the components as tight as you can with metal) and contribute an unnecessary additional fire hazard. Plastic reactor tanks are common in some off-the-shelf reactors because they are cheap; these are best avoided.

Sections of hose can be useful in your design, especially if it is transparent; however, ensure the hose you use is sufficiently heat-proof and reinforced so that it will not go soft and collapse on the vacuum side of your pump, or expand and split on the pressure side.

Secondhand hot water tanks have usually been discarded because there is something wrong with them; they may be split or have damaged ports or simply be full of lime scale. However, good ones do find their way into the recycler's yards and can be picked up for their scrap value. Other tanks can be found in the scrap yard; have a look and see what they have. Occasionally, ideal stainless steel vessels can be bought for their scrap value; otherwise, new tanks are ideal.

Simple processors

Possibly the simplest of home-built reactor consists of an old 205-liter (45 UK or 55 US gallons) oil drum with some paddles driven by an electric motor. All the components are cheap, or even free, and a conical bottom is sometimes welded onto the drum or its base, simply beaten out a little to form a convex base.

> It is not easy to make good-quality fuel in one of these. Plans for them are common in older biodiesel books but the design is very crude and fundamentally dangerous for the operator. It is easier to make a superior and safer reactor than to make one of these.

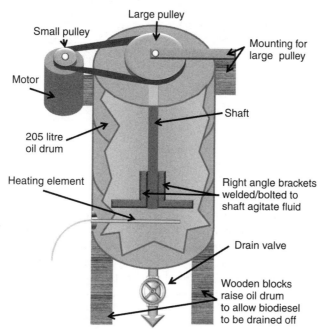

▲ Figure 7.1
An early simple design of reactor: not recommended.

Better designs

Better designs do not have an open top and use a powerful pump, rather than paddles, to mix the reaction.

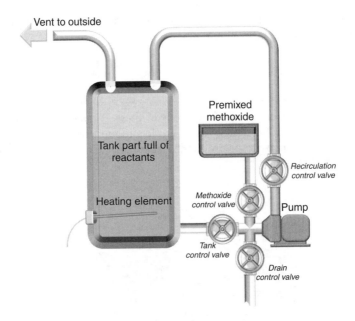

Vent to outside

Premixed
methoxide

Tank part full of
reactants

Recirculation
control valve

Heating element

Methoxide
control valve

Pump

Tank
control valve

Drain
control valve

▲ Figure 7.2
Basic design of a more sophisticated reactor.

The pump is multipurpose in this design; it is used to fill the reactor, to add the methoxide to the reaction, and to mix the reaction. This is achieved using multiple values, which are opened and closed to alter the route the oil takes. There is a heating element in the reaction vessel and a vent in the top; the vent is essential to ensure there is no chance of the vessel becoming pressurized or collapsing. It must terminate in a well-ventilated external space away from the operator as singnificant quantities of methanol vapor will escape from it.

The US Appleseed biodiesel reactor

The Appleseed reactor is very popular in the USA because of its simplicity, cheapness, and inbuilt safety. Developed by Maria Alovert (aka Girl Mark) from an array of predecessors, it is designed around the flat top and bottomed type of domestic water heating tank common in the USA and Canada; a typical 40-gallon water tank has space to make about 25 gallons of biodiesel.

The great thing about using a hot water tank, as opposed to other metal tanks or drums is that all the inlets and outlet ports, the heating element, and the thermostat are already there, as if designed for the job. The tools required to build it are only a screwdriver and a pipe wrench and some PTFE tape. No welding or other specialist skills are required, and you can buy just about all the pieces needed at any good plumbers or hardware store.

The methoxide mixing for this type of reactor is usually done using the carboy method, described below.

Polypropylene "cam-lever couplings" or "quick connects" (Figure 7.5) are an inexpensive way of making the process of swapping pipes about and connecting and disconnecting different bits of your processor much easier and quicker. Swap the hose tails in the processor diagrams for cam-leaver couplings, and put corresponding couplings on the ends of your hoses.

Using the illustrations you should be able to assemble a simple Appleseed, but there are virtually limitless different versions and adaptations to be found on the Internet.

Heating the oil in an Appleseed is done with the water-heater's built-in electrical heating element. Typically, there are two elements and two thermostats on this type of water heater. When used as a biodiesel processor, the top heater should be disconnected, as it may not be submerged in the oil, and the bottom heater rewired so that both thermostats are used in series, just in case one of them should fail.

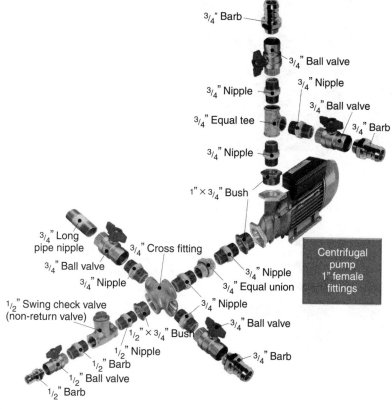

▲ Figure 7.3
Appleseed plumbing parts.

Figure 7.4
*Appleseed processor. (Courtesy of b100
supply.com.)*

Cam-lever couplings

Male
connector

Female
connector

Figure 7.5
*Cam-lever couplings, or quick connects, speed up
and simplify the process considerably.*

▲ Figure 7.6
Top of an Appleseed, note the pressure release and the condenser. (Courtesy of b100 supply.com.)

In early versions of the Appleseed, measuring the temperature of the contents of the reactor was done using an electrical car's engine temperature gauge; later versions use an analogue rotary dial-type thermometer with a threaded end that screws into the plumbing fittings and a probe that extends into the pipe work.

In early designs the tank was not vented, in order to stop valuable vaporized methanol from escaping, and allowing it to become slightly pressurized. Primarily because of safety concerns, it is now more common to have an open vent and to condense the vaporized methanol before it escapes. Condensers can be as simple as a vertical section of metal pipe a foot or two long, connecting the vent to the pipe leading to the outdoors, or more complex coiled or water-cooled condensers. Vents always terminate outside of the space you are working, as some methanol vapor will escape, especially when filling from empty. In addition to the open vent, it is also common to fit a pressure-relief valve, set to just a couple of bars, and also terminating outside in case something should block the open vent.

Pumps

The electric pump for your reactor is going to need to be a fairly chunky one in order to pump the heavy oil and ensure that enough mixing occurs. Beware of nonsparkproof motors; these are a hazard when working with potentially flammable and explosive materials like methanol (amazingly, some kits don't supply sparkproof motors!).

Lists of suitable pumps can be found on the Internet; 1-inch "Clear Water" pumps are popular, but your choice will depend on what country you live in, if you need a 110 V or a 230 V pump, and how big your processor is.

Figure 7.7
Nonpriming electric pump. (Courtesy of biodieselfilters.co.uk.)

Figure 7.8
Self-priming electric pump. (Courtesy of biodieselfilters.co.uk.)

Detailed information on how to build an Appleseed, its design variations, and a ton of other very good biodiesel information can be found in Maria's excellent self-published book, *The Biodiesel Homebrew Guide*, available from www.localb100.com. While its content is aimed at a US audience, the vast majority of the information is very applicable to anyone making biodiesel.

The best Appleseed tutorial on the web is on the Collaborative Biodiesel Tutorial site:

www.biodieselcommunity.org/appleseedprocessor/

And the excellent Infopop forum has a lot of information on the Appleseed in its Biodiesel Equipment section:

biodiesel.infopop.cc/eve/forums/a/frm/f/919605551

Or by searching Google for Appleseed.

UK hot water tank

The British variation of the Appleseed design uses the domed-top domestic hot water tank, very common in the UK, usually turned upside down so that the domed top becomes a pseudoconical bottom, which is better for a clean separation when draining the glycerol by-product layer from the biodiesel.

With this type of water heating tank, the electrical heating element is usually at the top of the tank; however, it is always submerged as the tank should always be full of water. When used as a biodiesel reactor, the tank is no longer used full. There is an air space at the top, but because it has been inverted the heating element still is always submerged.

When used as a reactor vessel, this tank has many of the advantages of its American equivalent but, unlike the US equivalent, the domed top tank, when inverted, does not come with an obvious port for use as a vent. It is essential that one be included to stop the tank from becoming overpressurized or from collapsing when cooling, and the vent terminated somewhere sensible.

Figure 7.9
Example of a British-style tank reactor. (Courtesy of Graham Laming.)

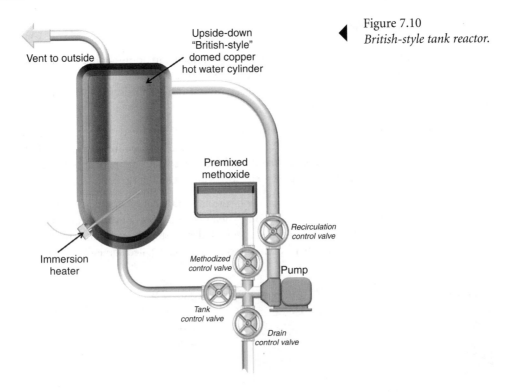

Figure 7.10
British-style tank reactor.

To add a vent, obtain a 15 mm copper tank connector and drill a hole in the center of the flat end of the tank. Of course, the end of the tank is not actually flat; like the flat top and bottom tank, it is actually "wine bottle bottom" shaped. Then, remove the electrical heater and, using a broomstick and Blutak if necessary, put the tank connector through the hole and do up the nut from the outside. For a very good seal, you could also braze it in place with a little solder. Using either hose or some copper pipe, run this vent to somewhere sensible outside.

Graham Laming's Eco-System

Graham Laming in the UK has developed the domed-top domestic hot water tank idea more than anyone and has come up with numerous enhancements, some bordering on genius, which could equally be adapted to work with any type of tank.

His "Eco-System" can dewater the WVO, process it into biodiesel, and recover the leftover methanol from the biodiesel. It also avoids water washing altogether. The design incorporates a venturi, which is used both to suck the methoxide into the reactor and to circulate all the vaporized methanol and water through a condenser for the dewatering and methanol recovery stages.

Graham's website has detailed information, helpful advice, and a load of other great ideas and the Infopop Forum has a detailed discussion of Graham's designs.

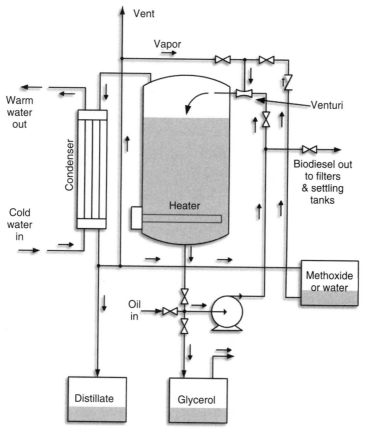

▲ Figure 7.11
Diagram of Graham's "Eco-System" processor.

Graham Laming's website has detailed information and instructions on his processor and a lot more too; very highly recommended: www.graham-laming.com

The Infopop Forum also has a lot of information and discussion on Graham's Eco-System processor in its Making Biodiesel section: biodiesel.infopop.cc/eve/forums/a/tpc/f/719605551/m/9921000191

How Graham's design works

Detailed instruction on how to build and run Graham's design is available on his website, graham-laming.com. A simplified description follows to aid understanding of his reactor design.

Initially, the tank is filled with prefiltered oil using the pump.

The oil is then circulated around the tank with the heater on to dry the oil. The venturi sucks in air, which first draws wet air through the condenser and dries the air before returning it back to the tank via the venturi, where it comes into contact with the wet oil and takes some of the moisture from it. This circulation carries on until the oil is hot and dry.

Next, the oil is allowed to cool, a sample is taken, and a titration is done to determine how much catalyst is required. The methoxide is mixed using the carboy method and the carboy attached to the reactor. With the pump running but the heater switched off, the valve is opened slowly so that the venturi can suck the methoxide from the carboy slowly.

As the tank fills up with methoxide, vapor and air will escape, displaced by the methoxide. Any methanol fumes will be caught in the distillate tank rather than escaping into the environment, so you can use us it another time.

Once all the methoxide is added, at least 20 minutes, the reaction is underway and the pump is left running, keeping it all well mixed. The venturi continues to circulate the vapor and catch any methanol, which should be very little. The reaction is allowed to run for up to 2 hours.

After the reaction has finished, the pump is switched off. The methoxide container is filled with water with about 1/20th of the oil you started with. Switch back on the pump and use the venturi to suck up the water. It is not important this time for it to be gradually introduced. Circulate it for about 15 minutes.

Turn everything off and allow the glycerine and by-products to settle out. Leave for about 90 minutes but not much longer or your glycerine may begin to turn solid in the pipes. Open the bottom valve and drain the glycerine.

Now turn your heater back on and run the pump. The biodiesel in your reactor is heated and circulated and any methanol and water it contains is evaporated and caught in the distillate trap.

The biodiesel is transferred to a settling tank, where, because we have removed the excess methanol from the biodiesel and dried it, the soap and any remaining glycerol will readily drop out.

Test the fuel before use, but it should be well converted, dry, and free from soap.

How to make a vapor condenser

Vapor condensers are easy to make. Water or methanol vapor enters the top of the condenser and travels slowly down the inner pipe while cold water flows upwards in a jacket around the inner pipe. As the vapor travels down, it comes into contact with the cold wall of the pipe, where it condenses into a liquid and runs out of the bottom.

This is a simple design. Many slight design improvements are available on the Internet to encourage a turbulent flow in the inner pipe and the outer pipe, increasing contact and, therefore, more efficient heat transfer; see Graham Laming's website or, indeed, moonshining websites.

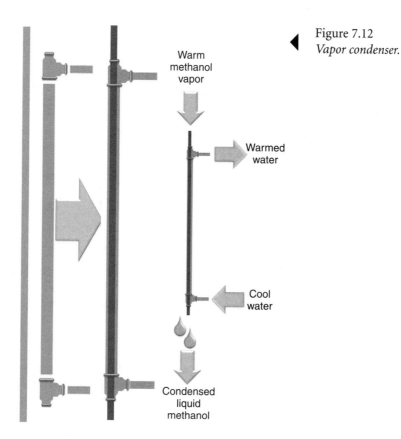

Figure 7.12
Vapor condenser.

Warm methanol vapor

Warmed water

Cool water

Condensed liquid methanol

You will need:

- A length of 15 mm and 22 mm copper pipe
- Two 15-15-22 copper T's
- Pipe cutter
- A blow lamp
- Some plumber's solder
- A little plumber's flux

Cut the 22 mm length shorter than the 15 mm length and solder the pieces together, as shown in Figure 7.13.

How to make a venturi

A simple venturi can be made from some copper plumbing parts soldered together. More complex designs can be found on the Internet, especially on Graham Laming's website and the Infopop Forum.

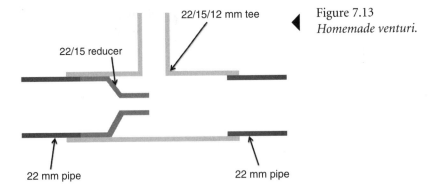

22/15/12 mm tee

22/15 reducer

Figure 7.13
Homemade venturi.

22 mm pipe

22 mm pipe

You will need:

- Some 15 mm and some 22 mm copper pipes
- One 22-15-22 T connector
- One 22-15 reducer
- A pipe cutter
- A blow lamp
- Some plumber's solder
- A little plumber's flux

Solder the pieces together as shown in Figure 7.13; you may need to make some alterations to the 22-15 reducer to get it all to fit together. With water flowing from left to right, you should get suction on the top 15 mm pipe.

Methoxide mixers

Dissolving your catalyst into your methanol needs to be done with care. As you should know by now, methanol is poisonous, almost odorless, and flammable, and the catalyst is caustic – it will burn your skin and is especially nasty if you get it in your eye. Nevertheless, you will need to measure significant quantities of both and devise a method of mixing them together while not endangering yourself or anyone around you. In addition to the hazards presented by these chemicals, the reaction between them is exothermic (it produces heat) and the catalyst can be reluctant to dissolve. You must never add undissolved catalyst to your reaction.

Carboy method

The carboy method of adding methoxide to the reaction is very common because of its simplicity. It uses any plastic carboy; 5 gallons Fort-Pak HDPE carboys sold by USA Plastics are commonly used.

It is usual to follow this procedure while your oil is heating inside the reactor.

- In a very well-ventilated area, preferably outside, you measure the methanol you need for your reaction into an HDPE "carboy" and put its lid back on.
- Then, still following all the usual safety procedures, weigh out the catalyst you will need, remembering to close the catalyst's container as soon as possible to minimize its exposure to the air.
- Back in the very well-ventilated area, take the carboy's lid off and, using a wide-mouthed funnel, add the catalyst to the methanol and put the lid back on.
- Label the carboy, so you know what it is it later.
- Check that the lid has sealed and gently rock the carboy for a few moments and then leave it alone for a while.
- Notice how the methanol gets warm by putting your hand on the outside of the carboy.
- Every 10–20 minutes, return to the carboy and give it another gentle mix. You are trying not to dissolve the catalyst too fast, which will cause the methanol to get very hot, while ensuring it does all dissolve eventually. Do it slowly and carefully the first time, maybe speeding up a little the next time round.
- Once all the catalyst has dissolved, and once the oil in the reactor is ready, you can add it to your reaction. This is usually done by making a special lid for the carboy, with a valve and a length of pipe to connect it to the reactor's plumbing. By adjusting the valves slowly, the reactor's pump can be made to suck in the methoxide while circulating the oil.

To make a carboy methoxide mixer you will need to take its cap and make a hole in it so a tank connector or similar can be attached through it so that a ball valve and a hose barb (or even better than a hose barb, a cam-leaver coupling) can be attached; see Figure 7.14.

The cap is removed only to put the methanol and the catalyst in the carboy. Be careful to avoid breathing in any fumes, especially if the carboy has been used before. Put the lid on as soon as possible and ensure the valve is closed and the lid does not leak before gently agitating the mixture to dissolve the catalyst.

Other methods of mixing the catalyst into methanol safely are as numerous as there are people to dream them up. Generally speaking, you are trying to ensure you have dissolved one in the other with the minimal risk to the operator; you are looking to be able to mix them in a sealed container by some means that won't cause a spark (so no electric motors) and that enables you check that it has all dissolved without opening the container.

Vents and methanol recovery

Having a vented reactor is very important. When filling the reactor, methanol fumes and air will need to escape; when emptying the reactor, air will need to enter it to replace the fuel that is leaving it. If the reactor was not vented then, while heating the oil, pressure could build up and when it was allowed to cool the opposite would happen and it would collapse.

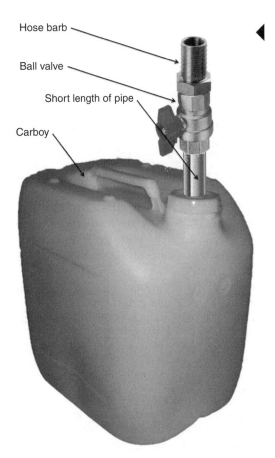

Hose barb

Ball valve

Short length of pipe

Carboy

Figure 7.14
Diagram of a carboy methoxide mixer.

In addition, empty tanks will contain a mixture of methanol vapor and air; if the heating element was accidentally switched on at this time, pressure could build up, possibly resulting in an explosion.

This said, many people open and close their vent at different points in the reaction process to stop valuable methanol escaping, while others don't like this as it is too easy to accidentally leave it closed.

Your vent must always lead to somewhere where the methanol vapor can safely escape, typically via a hose to the outside. Remember, methanol is heavier than air and will sit on the floor and collect in hollows.

Recovering leftover methanol

In the biodiesel reaction, methanol tends to be added at about 20–25% by volume. Only about half of this is used up in the reaction, the other half is used to drive the reaction in the

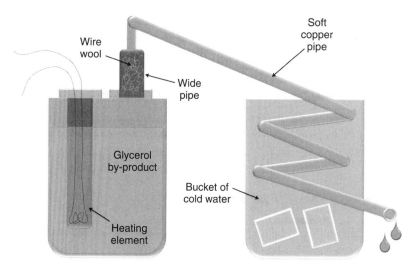

▲ Figure 7.15
Simple methanol recovery still.

right direction, from triglyceride to FAME, and to keep conversion rates high. What we glibly refer to as glycerine or the glycerine by-product will only be about 40% glycerine; roughly 25% of it is methanol, along with some catalyst and soap and other impurities.

The methanol in the glycerine by-product is recoverable in a still, and this may be worth doing as methanol is likely to be the largest cost in your fuel's production.

See Chapter 12 on what to do with your glycerine by-product.

Figure 7.15 is a diagram of a simple still; another one is described earlier in the chapter for recovering methanol from the vapor escaping from your reaction; more complex still designs can be found on the Internet, primarily on moonshine how-to websites! However, while the apparatus is the same, methanol is one type of alcohol you do not want to be drinking!

Methanol boils at around 60°C (140°F). You will need to try to hold your still's temperature around 55°C to (131°F) to avoid also boiling off water or other impurities and reducing the purity of your recovered methanol.

You can use a hydrometer to assess the purity of the methanol you produce.

Settlement tanks and wash tanks

Unless you are using new oil, you will want to do some prereaction processing to remove the larger particles of food from your used oil in a settlement tank and dewater your oil before it is made into biodiesel.

After the reaction has completed, you will need to separate the biodiesel from the glycerol by-product (this is more commonly done in the reaction vessel, but not always) and further process it by washing and drying it before it is filtered, stored, and ultimately burned in your engine.

Details on these processes can be found in Chapters 6 and 8 of this book.

There are many designs for settlement tanks and wash tanks from the off-the-shelf conical bottomed plastic ones, to making your own metal conical bottom tank, to the supersimple "standpipe" design.

How to make a standpipe tank

Standpipe tanks are simple to make, very common, and very effective as settlement tanks and wash tanks. They can also be used to make dewatering tanks, drying tanks, and even storage tanks.

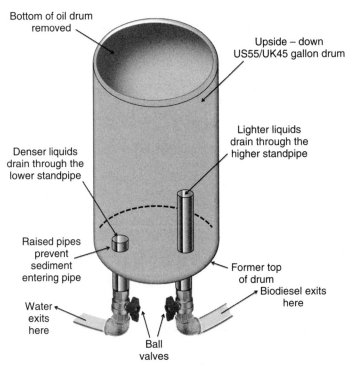

▲　Figure 7.16
Diagram of standpipe tank.

They are usually made from old 205 liter (45 UK or 55 US gallons) oil drums, the type with two threaded ports in the top, not the ones with a removable top.

The drum is inverted and its bottom, now its top, since it is inverted, is cut out: a jigsaw is ideal for this. Don't forget to debur the edge with a file. It is not a good idea to leave the tanks uncovered, however, as dirt or small mammals can easily fall in.

The two ports in the top, now the bottom since the tank is used upside down, are fitted with off-the-shelf plumbing parts in such a way that one drains directly from the bottom of the tank while the other drains from near the bottom. By selecting which you drain the tank through, you can separate water from biodiesel or glycerol from biodiesel or whatever.

Once you have cut out the top, what used to be the bottom of the drum, you will need to assemble the following using lots of pipe tape as you go:

- One ¾-inch BSP to 2-inch BSP bush: Put this into the 2-inch hole; you now have two ¾-inch holes.
- Two ¾-inch BSP nipples, about 2-inches long: Screw these into your two ¾-inch holes.
- Two ¾-inch BSP 45° elbows: Screw these to the nipples.
- Two ¾-inch BSP pipes long enough to protrude from under the drum: Screw these into your 45° elbows.
- Two ¾-inch valves: Screw these into your protruding pieces of pipe.
- Two ¾-inch hose barb adaptors: Screw these into the valves.
- Enough hose to get from your standpipe tank to wherever you want to drain the fluids to. You probably want to use clear hose so you can see what is coming out.
- One piece of ¾-inch pipe, about one-quarter of the height of your tank: See note below; screw this into the inside of the ¾-inch BSP to 2-inch BSP bush.
- A suitable stand that is strong enough to support several hundred kilos, while not putting pressure on the plumbing, and high enough off the ground to be useful.

Note: The length of the standpipe depends on a few things but will typically be about one-quarter the height of the tank. Obviously, in order to draw the less-dense liquid from the top without also drawing the bottom liquid, the standpipe will need to pass all the way through the bottom liquid and into the liquid above; you may need to adjust the length of the pipe accordingly.

The parts for the standpipe tank can be bought from most hardware stores or plumbers' merchants, or bought as a kit from many biodieselers' suppliers.

Standpipe settlement tank

The standard standpipe tank makes an ideal settlement tank. Pour in your feedstock and leave it to stand for as long as you can. The bigger particles will sink to the bottom and the cleaner oil will sit above it. Cleaner oil can be drawn off using the standpipe valve while dirtier oil can be drained off later using the other valve.

Standpipe dewatering tank

If you are using the standpipe design as a dewatering tank you will need to add insulation and a heater to the above design. The length of the standpipe will depend on how much wet oil you are processing. Wetter oil will typically be about 15% of your oil. The procedure for dewatering WVO is described in detail in Chapter 9. After leaving the oil to stand, the dryer oil can be drawn off using the standpipe valve and afterwards the wetter oil drained using the other valve.

Standpipe wash tank

The standpipe tank makes an excellent wash tank. Fill your standpipe tank with your biodiesel and then add enough water to nearly come over the standpipe. Pump the water from the nonstandpipe valve and, using a mister or similar, spray it back over the top of the biodiesel. Detailed instruction on mist washing can be found in Chapter 6. Afterwards, you can either drain the water and replace it with more water for further washing, or drain the biodiesel instead using the standpipe valve, leaving the last wash's water still in the tank ready to be used with your next biodiesel batch. As it was the water from the last batch's last wash, it should not be heavily contaminated and so completely usable for your next batch's first wash.

Standpipe drying tanks as storage tanks

It is not necessary to use a standpipe tank as a drying tank or short-term storage tank, but they do make adequate ones.

To dry your biodiesel in a standpipe tank, either simply leave your fuel to stand somewhere well ventilated and the water will "dry" from it, or actively speed up the process by pumping the fuel out of the bottom and into the top while actively ventilating with a fan.

Standpipe tanks can also be used to store your biodiesel for a short while, but they are not suitable for long-term storage. Just make sure it is well covered and that your fuel is filtered to 10μm or better before use in case it picked up dirt while it was in the tank.

 The Collaborative Biodiesel Tutorial site has information on Standpipe design: www.biodieselcommunity.org/standpipewashtank/

Plans, kits, and readymade reactors

Of course you don't have to design, source, and build your own processor. There are hundreds of companies who will sell you anything from a set of plans to a fully assembled automated reactor. Below are just a few of the reactors available to buy, either as plans or kits or fully built machines, in the USA, UK, and across the world. As you would expect, not all are created equal. The ones featured are examples, not necessarily endorsed or recommended.

Off-the-shelf plans and kits

Some processor plans for sale are well-developed, well-thought through designs based on real reactors with a well-researched and sourced parts list; others are simply copied plans from the Internet that you could get for free from 5 minutes independent research.

Off-the-shelf Appleseeds

Appleseed reactors can be purchased as a kit for not much more than the cost of the parts and saving you the bother of sourcing them.

The Appleseed Biodiesel Processor Kit from B100 Supply: www.b100supply.com

Twyn Tub

▲ Figure 7.17
The Welsh company Goat Industries has its "Twyn Tub" 100 and 150 biodiesel processors available as either plans, a kit, or an entirely built machine. (Courtesy of Patrick Whetman, Goat Industries, www.vegetableoildiesel.co.uk/twyntub100.html.)

Dynadroit biodiesel

Figure 7.18
Dynadroit steel biodiesel processors. (Courtesy of Dynadroit Biodiesel: www.dynadroitbiodiesel. com/biodiesel.html.)

Murphys machines

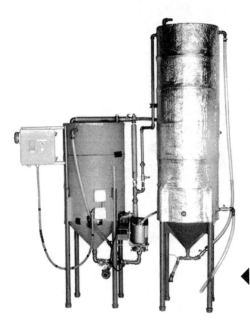

Figure 7.19
All steel modular biodiesel processor from Murphys Machines. (Courtesy of Murphys Machines: www.murphysmachines.com.)

Fully automated reactors

AGR Energy manufacture very high-quality reactors for the USA market that you simply fill with raw ingredients and walk away while it makes biodiesel for you. Designed for minimal operator interaction, they have a range of fully automated stainless steel reactors; it does it all for you from used oil through dewatering, reacting, separating, washing, and drying, to high-quality biodiesel; from start to finish, about 60 hours with minimal supervision. They even offer a big discount when you trade in your old plastic reactor!

◄ Figure 7.20
AGR Energy's BioPro 150. (Courtesy of SpringBoard Biodiesel: www. agrenergy.com.)

There are a number of cheap plastic reactors available, especially in the USA, marketed at the novice biodieseler. Their price is very tempting but they have a terrible reputation in the biodiesel community. Their instructions are often poor quality; they have a tendency to produce low-quality incompletely reacted biodiesel; and they expose the operator to a number of very significant dangers.

The choice of plastic components in their design appears to be motivated by price and profit.

They use a similar principle of mixing the reaction with a pump, similar to other designs, but they don't include a heater as, obviously, heaters and plastic vessels don't go together. Instead you either have to process the oil cold or heat it away from the processor and then transfer the hot oil into the reactor for processing, where it will cool quite rapidly.

When oil is added to some of these reactors above about 40°C they have a tendency to leak. It is possible to get a good conversion ratio at lower temperatures then 50°C (the rule of thumb is to increase the mixing/reaction time by 1 hour for every 10° below 50°). But this will increase your reaction time quite significantly.

Methanol and catalyst mixing in these reactors is also very crude and often results in the catalyst dissolving incompletely. This means the operator has to open the methoxide tank, exposing himself to and releasing the methanol fumes, and mix it by hand.

An alarming number of plastic reactors have caught fire, thus spilling burning fuel. While usually because of operator error, this would have been avoided if the reaction vessel had not been plastic.

While some plastic reactors are indeed not as bad as others, they all suffer from the same primary design flaw: They are built around a plastic tank. For less than the price of one of these plastic reactors, you can buy a complete Appleseed kit, which is superior in every respect.

Chapter 8

Engine modification – running on SVO

So, you've read Chapter 6 on making biodiesel, and it all seems like a bit too much trouble – you'd much rather just put the vegetable oil in your car and drive off? Read on! In this chapter, we're going to discuss converting your car to run on straight vegetable oil fuels that have not gone through the process of any chemical wizardy. Oil in, drive off ... simple!

As we have discussed previously, the principal issue with getting the diesel engine to run on vegetable oil is viscosity.

In Chapter 6, we used transesterification to chemically modify the vegetable oil and make it more like mineral diesel: we were modifying the fuel to make it more compatible with the engine.

In this chapter we are talking about running diesel engines on straight vegetable oil, SVO (or pure plant oil, PPO, or whatever you want to call it); unmodified, unblended, and unused vegetable oil. In order to do this we will modify the engine to make it more compatible with the fuel.

Some people take this one stage further and run their modified engines on waste vegetable oil, WVO (often referred to as used cooking oil, UCO, or used vegetable oil, UVO). The information in this chapter applies to this too; additional information on running on WVO can be found in the other chapters.

Modern diesel engines are designed to run on mineral diesel; some are very intolerant of other fuels. Some older diesel engines, most notably some 1980s Mercedes and older IDI engines, are remarkably tolerant of a wider range of fuel types and viscosities and have been known to run unmodified, supposedly without any problems, on other diesellike fuels, including vegetable oil; however, for the vast majority of us, we need to either modify the fuel or the engine.

There are also many people who claim to run their unconverted engines on all manner of things. In the same way that we all know smoking cigarettes causes cancer and yet know of a guy who smoked 40 a day and lived until he was 100 years old, running your unconverted car on vegetable oil will damage it, even if some people appear to get away with it.

Just about all diesel engines can be converted to run on vegetable oil. All types of injection systems can be modified; it is just a case of choosing the best system for your engine. There are far too many different engines and engine types for there to be a complete answer for everyone here; you will need to do further research for your engine.

The basics

In this approach to running your diesel engine on biofuel, rather than modifying the fuel to make it more compatible with the engine, as in biodiesel, the engine and fuel system are modified to make them more compatible with the fuel.

Depending on the type of oil and the ambient temperature, vegetable oils are between 11 and 17 times more viscous than mineral diesel. If we heat the oil, it becomes runnier and therefore more diesellike. Try putting a little cold oil in a pan and gently heating it while rocking the pan from side to side and you will get a feel for how the temperature of the oil affects its viscosity.

If the oil is not sufficiently diesellike, it will greatly increase the strain on the engine's fuel system and when it is sprayed into the combustion chamber it may not atomize properly, resulting in incomplete combustion, coking, poor performance, higher emissions, and ultimately engine failure. At around 70°C (160°F) the viscosity of SVO becomes comparable to mineral diesel and almost any diesel engine ought to run very happily on it.

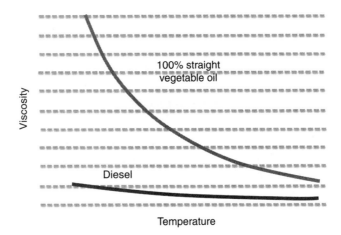

▲ Figure 8.1
Graph illustrating the relationship between temperature and viscosity for mineral oil diesel and SVO.

Besides viscosity, vegetable oils obviously have other different properties to mineral diesel. They are chemically different and have different combustion characteristics in that vegetable oils have a higher flash point and they ignite less readily and have a lower cetane number.

Heating the fuel

Internal combustion engines are inefficient things and the diesel engine is no exception. On average even a very efficient engine will only achieve about 20% efficiency: i.e., 20% of the fuel's energy goes to the wheels to drive you forward and 80% is lost, primarily as heat. We put some of this heat to useful work when the weather is cold to heat the vehicle's cabin, but most of it is just dumped into the engine's cooling system.

Heat exchangers work by taking a little of this waste heat and using it to heat the fuel. Purpose-built units designed for keeping mineral diesel flowing in cold climates can also be used very effectively for getting SVO up to a temperature where its viscosity is similar to the mineral diesel.

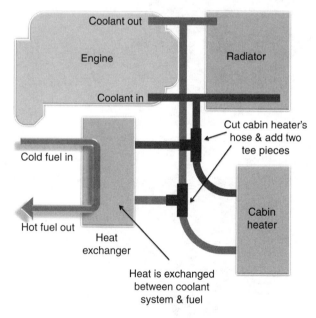

▲   Figure 8.2
How a heat exchanger fits into existing cooling circuit.

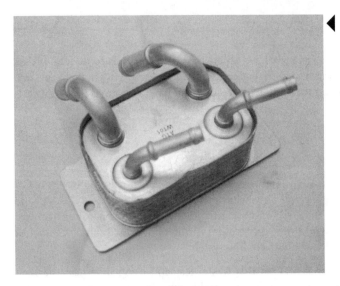

Figure 8.3
Heat exchanger from an ATG kit. The ATG kit also comes with a supplementary electrical heater to ensure the oil is up to temperature.

Figure 8.4
Heat exchanger from an Elsbett kit.

Figure 8.5
Heat tee pieces and hose clamps. These are used to "tee into" the cabin heater's hoses. Ensure you buy the correct size for your engine's hoses.

How to make a heat exchanger

Heat exchangers can be bought from SVO conversion suppliers and engineering suppliers but you may want to have a go at making your own. It's not a particularly difficult task if you've got a smattering of knowledge regarding working with metal, or have ever done a bit of household plumbing before.

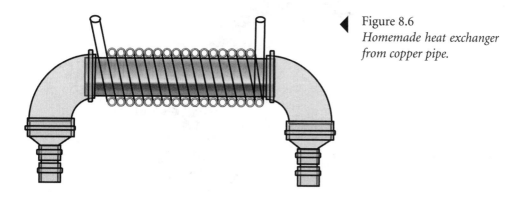

Figure 8.6
Homemade heat exchanger from copper pipe.

You will need:

- A length of soft copper pipe, diameter to match your vehicle's fuel pipe diameter, typically 8 mm
- A short length of wide copper pipe, 22 mm or larger
- Copper plumbing fittings to enable you to fit the wide copper pipe to your cabin heater circuit
- Some fine sand
- Plumber's solder
- Plumbing flux
- Wire wool
- Blow lamp

This simple design is made from some soft copper pipe wrapped around a larger piece of copper pipe; the wide copper pipe is plumbed into the hoses leading to your vehicle's cabin heater while the narrower pipe goes into your SVO fuel line.

All copper parts must be thoroughly cleaned with wire wool before you start. First, fill a length of the narrow soft copper pipe with sand; then, wrap it around the wider pipe. The sand inside it will stop its walls collapsing or getting kinked. Take care to make a neat job of it, with each turn sitting tightly up against the last and tightly around the wide copper pipe. Once it is tightly curled around, pour out the sand and solder the two pieces together using the flux, solder, and blow lamp.

If you have never soldered copper before, we recommend getting someone to teach you and have a few practice goes on bits of copper pipe before attempting to make the heat exchanger.

You can also make a hose-in-a-hose heat exchanger where the fuel pipe is run for a short while inside a hose with hot water from the engine in it.

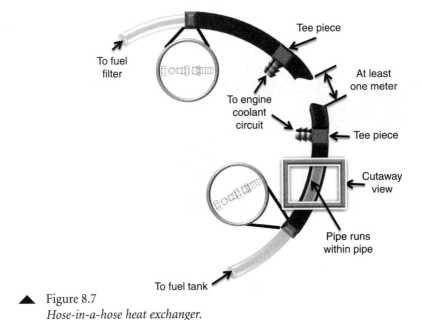

Figure 8.7
Hose-in-a-hose heat exchanger.

 The antifreeze which is added to the coolant water in your engine is both poisonous and sweet tasting. Don't let it spill in the ground or collect in open containers, as dogs and cats may well take it upon themselves to drink it, and it won't do them any good at all.

Filters

Most vehicles have at least two separate stages of filtration: There is a screen in the fuel tank to trap large pieces and stop them blocking the fuel lines, and then another element, usually paper, to stop small particles entering the engine. They are rated, in microns (µm), by the size of particle they stop; 1 micron is one-millionth of a meter,

1×10^{-6} m (1 μm). The final filtration on most engines is about 10 microns (μm), but some go better.

It is also common for a water trap to be incorporated into the final filter, with a drain plug at the bottom, and for the return lines from the engine to the tank to also pass through the filter, which can be somewhat confusing.

Figure 8.8
Filter and its bracket from an Elsbett kit.

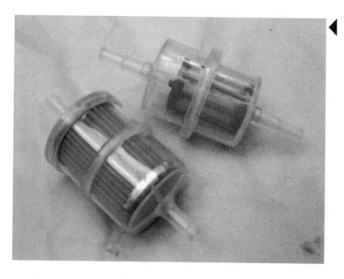

Figure 8.9
Prefilter and main filter from an ATG kit.

Racor filters

Racor have been making filters, water separators, and fuel heaters for over 35 years. They have established themselves with a reputation for making the best filters available and their prices match. They make filters down to 2 microns, heated filter housings, and fuel heaters for trucks and boats that are also very suitable for SVO use.

Cold weather

When SVO gets cold, it will freeze into a gel-like form that your engine's pump won't be able to move; at what temperature that this happens is dependent on the oil. Adding 10% diesel to your SVO is a very effective way of dealing with this problem: Just make sure you do it before the cold weather arrives!

Keep a sample of your SVO in a small clear plastic bottle in your car, maybe in the driver's door pocket. That way you can assess how frozen the SVO in your tank probably is.

You will find that there is a narrow range of temperatures where your SVO (and biodiesel) will begin to form little frozen crystals but will not completely freeze into a gel; while your engine's pump can probably still move it about, your fuel filter will become blocked by what is otherwise perfectly good fuel.

Whereas the adding of 10% winter diesel trick still applies to this problem, there are heat exchanger fuel-filter combinations available, both after market and fitted as standard on some vehicles, which also solves this problem. There are electrical heater jackets available especially designed to fit around the fuel filter and you could always make your own (see below).

None of these should be used as the primary means of getting your SVO up to temperature. A proper heat exchanger should always be used, but all can be helpful in ensuring frozen bits of fuel don't block the filter.

Figure 8.10
Electrical filter heater jacket to fit around the filter, supplied with an Elsbett kit.

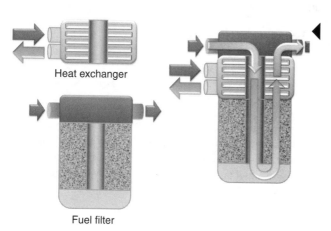

Heat exchanger

Fuel filter

Figure 8.11
Diagram of after-market heated fuel filter.

Homemade heated filter

A very similar method to the homemade heat exchanger can be used to make a heater for the cartridge-type fuel filter. This time, using 12 mm soft copper pipe, fill it with sand and curl it around an old filter cartridge the same diameter as the filter on your vehicle, again having cleaned it before you started with wire wool and taking a great deal of care to make a neat job of it, with each turn sitting tightly up against the last and tightly around the filter. Slide the spiral off the filter and, being very sparing with the solder, solder it into one piece by attaching the turns together. This can be plumed into the cabin heater circuit along with the main heat exchanger to warm the fuel as it passes through the filter.

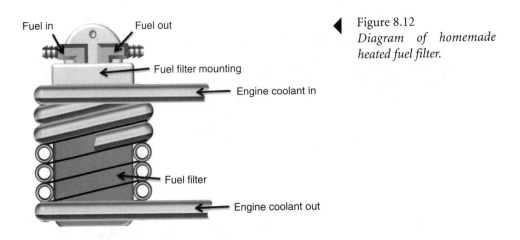

Figure 8.12
Diagram of homemade heated fuel filter.

Heated fuel tanks

Some people, especially those in colder climates, build heated fuel lines all the way back to the fuel tank and install a heater in the tank too. Some have experienced problems where these systems have leaked coolant water into fuel tanks, but, if done carefully this, should not happen.

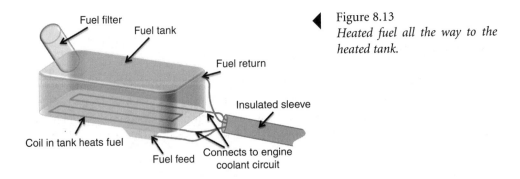

Figure 8.13
Heated fuel all the way to the heated tank.

Lift pumps

Lift pumps (also called supply pumps or feed pumps) are fitted between the fuel tank and the injector pump on many engines; their purpose is to ensure a good supply of fuel to the injector pump. Most are driven directly by the engine itself, but more recently vehicles come fitted with electrical ones and many SVO kits come with an additional electric one.

Additional lift pumps can be added to SVO systems if required. Some kits are supplied with them to help the vehicle's existing lift pump with the extra strain of moving the more viscous vegetable oil.

Injection systems

Indirect-injection engines (IDI) seem to be the most tolerant of SVO conversions. Direct-injection engines (DI) are less tolerant but generally convertible. Modern engines with common-rail injection systems should be converted to SVO with the utmost caution, maybe only with a high-quality kit. However, biodiesel is a better option.

For more information on injection systems, see Chapter 3.

Injection pumps

Inline injection pumps such as most Bosch models are most suitable for SVO. Other suitable manufactures include Diesel-Kiki, Nippon Denso, and Zexel.

There is considerable evidence to suggest that unsuitable engines include those with rotary injector pumps, including many otherwise convertible engines. Injection pumps made by CAV, Lucas, Stanadyne, RotoDiesel, or Delphi are particularly unreliable when used with SVO. While there are numerous examples of exceptions to this, rotary injection pumps should be avoided.

Engine oil

Under normal operating conditions a little diesel fuel can get into your engine's sump, and therefore the engine's lubrication oil. Older engines tend to suffer from this more than newer ones, but it does not matter as diesel fuel is volatile and so vaporizes in the hot engine and is burned off. When running on SVO the same thing happens, the fuel gets into the sump oil, but SVO is not as volatile as diesel and so tends to remains in the sump and this can cause problems. Keep an eye on your engine's oil level, especially when new to SVO, and watch for unusual changes in its level, especially if it starts to increase! A lot of SVO users simply change their engine's oil more regularly, some as often as every 5,000 miles.

Common rail

Today's modern DI engines are very highly sophisticated machines. The pressures on manufacturers to produce cleaner and more efficient vehicles have resulted in very clever computer-controlled systems that use fuel pressures in excess of 1500 bar, far higher than in more conventional engines. They inject precise quantities of fuel in droplets far smaller than older engines, even controlling individual burns with multiple injections. Fuel viscosity and combustion characteristics need to be well known for this type of technology to work properly, and using SVO in it may be too risky. Owners of common rail engines are better off looking only at the highest-quality kits designed specifically for your engine, or maybe consider going the biodiesel route instead.

One tank or two?

There are two main approaches to running on SVO: the two-tank and the one-tank conversion. In a one-tank system, the engine is converted to run directly on vegetable oil; in the two-tank system, the engine is started on mineral diesel (or indeed biodiesel) and only runs on SVO when the engine is hot. There is no "one solution fits all" approach; rather, each engine's conversion ought to be approached individually.

Common to both methods is the idea of heating the oil to reduce its viscosity, so a heater is added in the fuel system (often two or more) that can be either electric or take waste heat from the engine's cooling system.

Most engines come with at least a 10-micron filter, many as small as 1 micron. The tiny frozen particles in cold fuel can block your filter. One solution to this problem is swapping the original filter for one that allows the cloudy fuel through; some kits come supplied like this. This will certainly cure the problem, but will also allow particles of dirt through, which are sufficiently large to damage your engine. Fit a heater to your filter or use an antifreezing additive instead.

Both DI and IDI engines can be converted using the twin-tank method. Some engines may also be suitable for a single-tank conversion.

Two-tank systems

The usual approach to running on SVO is the two-tank or twin-tank system. This is where a second fuel tank is fitted to the vehicle along with a way of switching between the two. The engine is started on mineral diesel, or indeed biodiesel, and allowed to run for a while on this until it has heated up sufficiently for the heat in the engine's cooling system to be used to heat the vegetable oil.

At this point, either automatically or manually, the engine is switched over to the second fuel tank, which contains the vegetable oil. The heat of the engine is used to keep the fuel flowing and arriving at the engine sufficiently diesel-like and runny. When the driver begins to approach his destination, he switches from the vegetable oil tank back to the diesel tank and the engine is run for a while on the diesel so that, when it is switched off, and subsequently restarted from the cold, the engine is full of diesel and not cold vegetable oil.

It is a good idea to incorporate an alarm or buzzer that sounds when you switch the engine off to tell you if you have forgotten to switch back from SVO to diesel.

In some systems the vegetable oil is heated in the tank and in the fuel lines all the way to the engine. In other systems the fuel is heated only immediately prior to entering the injection pump. Some systems rely entirely on the engine's cooling fluid to be hot enough, whereas others use additional electrical heaters controlled by the temperature of the fuel.

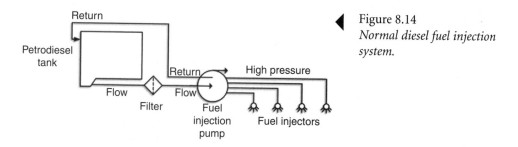

Figure 8.14
Normal diesel fuel injection system.

In the standard diesel fuel system, fuel is pumped from the fuel tank to the injector pump, where it is delivered to the injectors. The injector pump supplies a little more fuel than is required by the injectors, so small amounts of fuel are left over and fed back to the tank.

▼   Figure 8.15
    How to read these diagrams.

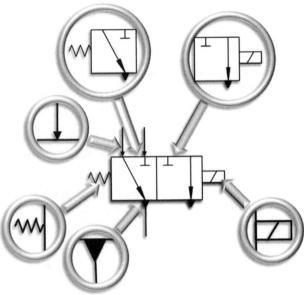

On pages 135 and 136 we look at the mechanical operation of a solenoid valve; however, engineering convention dictates that we write symbols out in a shorter, annotated form, to make things quicker.

In this book, we've tried to use the standard symbols for solenoid valve gear, as they provide a bit more information than just a "black box" drawing of the valve. However, if you're unfamiliar with these symbols, or have never encountered them before, here is a quick guide, so you know what to expect. Solenoid valve symbols consist of two or more adjacent "blocks." The "block" indicates the position that a valve can be in – i.e., "activated" and "deactivated" – you may encounter a symbol with "three" blocks; in this case, the middle block usually indicates the position that the valve "passes through" while switching from one mode to the other.

You will notice inside the block that there is a combination or "arrows" and "tees." The little "tee" that looks as if it is "capping off" a line, is doing exactly that; it tells you that when the valve is in this position, that particular route is sealed off, and no fluid can flow through it.

The arrow inside the valve indicates what port the opposite port is connected to, forming a path between one side of the block and the other.

You will notice around the perimeter of the blocks that there are some small "black triangles" that indicate that the port is an "output" for fluid (in this case vegetable oil or biodiesel). Note that if you see a diagram and these squares are white, shown in outline only with no fill, the valve has been designed as a pneumatic valve to carry air not fluid. On either side of the block, you will see a small zig-zag shape, and a rectangle with a line through on the opposite end of the symbol. This indicates the position of the solenoid and the return spring. Imagine the solenoid pressing against the spring when the valve is activated, "pushing" the block that sits next to the ports out of the way, and pushing the "alternative" block into its position. If you look to Figure 8.15, you can see the key points from this box highlighted.

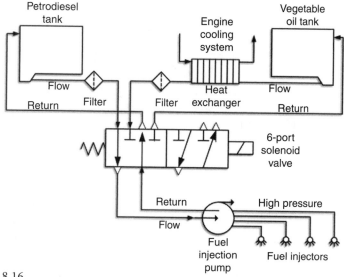

▲ Figure 8.16
Open-loop two-tank system with single six-port solenoid valve.

The open-loop two-tank system with single three-port solenoid valve is similar to the normal fuel system, but has been modified to include a second tank. Both flow and return are kept separate, but the driver can switch between the two fuel tanks; excess fuel is returned to the same tank it was drawn from.

It uses a single three-port solenoid valve.

As in the open-loop two-tank system, in the closed-loop two-tank system with six-port solenoid valve fuel is pumped from either one of two tanks selected by the driver. However, the system has been modified so that excess fuel from the injectors is not returned to a tank but straight back into the injection pump. This has the advantage of not wasting heat by dumping hot oil

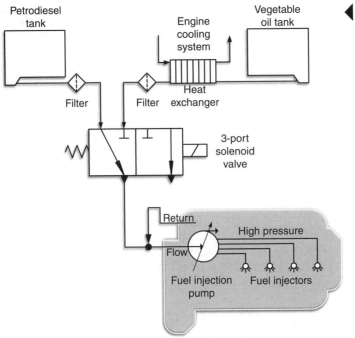

Figure 8.17
Closed-loop two-tank system with three-port solenoid valve.

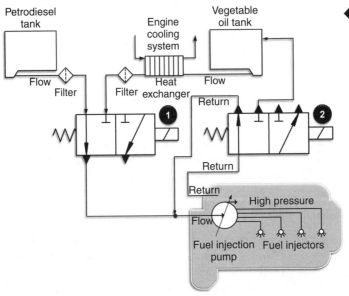

Figure 8.18
Closed-loop two-tank system with purge, double three-port solenoid valve.

back into the tank, increasing the chances of oil being supplied to the engine being of the desired temperature, and lessening the strain on the lift pump. Unfortunately, this system does not allow trapped air to escape from the system.

The closed-loop two-tank system with purge, three-port solenoid valve is a combination of the open-loop and the closed-loop systems, the best of both worlds. It functions exactly as the closed-loop system, except that when the driver switches from the SVO tank back to the diesel tank, the second, return valve is automatically actuated for a few seconds, turning the system from a closed-loop into an open-loop system for a few moments. This both allows any trapped air to escape and purges the injector pump of most of its vegetable oil.

Valves

Switching between tanks is done using electrical valves, either solenoid or motorized, connected to a switch on the dashboard usually via some fuses and relays. The solenoid valve is a simple affair. An electrical actuator, usually a solenoid, acts against a spring. In the solenoid valves that we are using, we don't simply want to stop and start the flow of a fluid, but we want to switch between two alternative paths (in this case, fuel to or from the vegetable oil tank, or diesel tank). In the case of the six-port valve, we want to make two changes – both from either tank, with the return going back to either tank. In this case, we use a solenoid valve that is either mechanically coupled – where there is a physical linkage between two separate valves operated by the same actuator – or, alternatively, we can use two separate three-port valves that are coupled electrically to achieve the same effect.

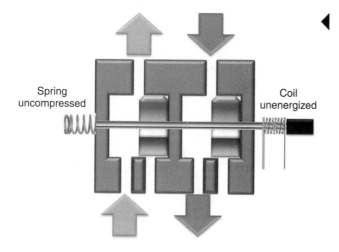

Figure 8.19
Diagram of a 6-port valve with unenergized solenoid.

Spring
uncompressed

Coil
unenergized

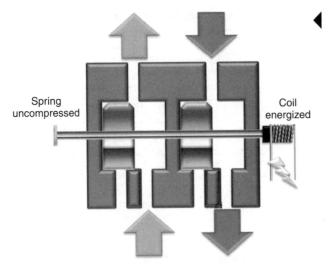

Figure 8.20
Diagram of a 6-port valve with energized solenoid.

Spring uncompressed

Coil energized

Figure 8.21
Pollak six-port solenoid valve.

Pollak makes very popular and relatively inexpensive valves in either 3-port or 6-port configurations, especially designed for switching between fuel tanks.

You can find Pollak valves here:

pollak.thomasnet.com/category/fuel-tank-selector-valves
www.stoneridge.com/

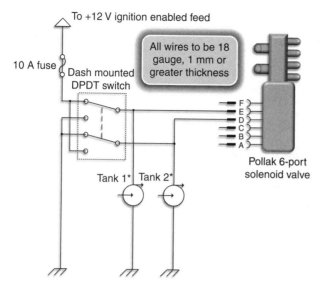

To +12 V ignition enabled feed

10 A fuse

Dash mounted DPDT switch

All wires to be 18 gauge, 1 mm or greater thickness

F
E
D
C
B
A

Pollak 6-port solenoid valve

Tank 1* Tank 2*

*Optional electrical lift pumps for low-pressure fuel pumping from tank to injector pump

Figure 8.22
How to wire up a Pollak 6-port valve to switch-on dash. Note, optional electrical feeds to separate lift pumps from each tank.

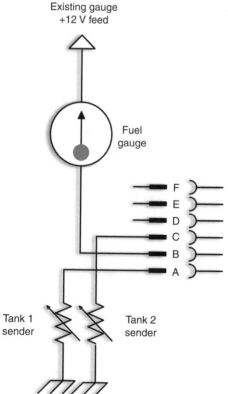

Existing gauge
+12 V feed

Fuel gauge

F
E
D
C
B
A

Tank 1 sender

Tank 2 sender

Figure 8.23
How to wire up a Pollak 6-port valve to feed the fuel gauge from the tank in use. Note that unless your fuel tanks and fuel tank senders are identical, you may need some additional components to make it read accurately for both tanks.

Figure 8.24
The two solenoid valves that come with the ATG two-tank kits are of exceptionally high quality. Note also the temperature sensor and supplementary "Dieseltherm" electrical heater visible in this picture, which come as part of the ATG kit.

Bleeding

Unlike a conventional petrol engine, diesel engines are not at all tolerant of having air in the fuel system. Making any alterations to the fuel system, even just changing a fuel filter, will inevitably cause some air to enter the system and you will need to get it out.

Unlike fuel oil, air is easily compressible; if there is any in the injector pump or the injectors, it will absorb all the pressure exerted on the fuel and the injector will not open.

You have to pump fuel through the system and loosen some of the parts of the fuel supply and then the injection system to get all the air out. The procedure for this is referred to as bleeding, and varies from engine to engine, and from vehicle to vehicle. If you don't know how to do this, you will either need to refer to a workshop manual or get some help.

The injection pump produces some very high pressures. While bleeding your engine, you may need to release some of the high pressure unions, potentially exposing you to fuel that is under enough pressure to penetrate your skin. Many mechanics have oily tattoos due to this, and it can cause blood poisoning.

Never be tempted to bleed your engine by towing it about while in gear. This will damage your engine. Bleed your fuel system properly to avoid irreparable damage.

For a two-tank system you will have to bleed both tanks. If you have modified your fuel system to a closed-loop system it will be especially hard to get the air out. If you have a closed-loop-with-purge you will need to hold the purge valve open while you bleed.

Advantages and disadvantages of the twin tank

The main advantage of this system is clear: You have the best of all worlds – the choice of mineral diesel, biodiesel, new vegetable oil, and waste oil – in just about any diesel engine.

The kits tend to be a little cheaper, and the conversion a little easier than a one-tank kit, as you are only altering the engine's fuel system and not the engine itself. The conversion is reversible and the kit can generally be transferred to another vehicle.

The main disadvantages of this system is that you have two tanks to worry about. You have to remember to switch over to the vegetable oil and remember to switch back to diesel again before you reach your destination. You can only switch to the vegetable oil tank when the engine is warm, which means that for around-town driving you may not use the vegetable oil as much as you might like.

You lose some space in the vehicle to the additional tank. Some people install a small additional tank for the diesel which does not take up too much room and can be fitted in a corner or even in the spare wheel well (though we have never worked out what you are then supposed to do with the spare wheel) and put the vegetable oil in the original tank.

In larger vehicles, the second tank can be as large as, or indeed larger than, the original tank, but a small one is still an option too. Which tank you then decide to put the SVO in is up to you, but it is an important decision to make before you carry out the conversion, as changing your mind later may involve a significant amount of work. From a refilling point of view, it is often easier to keep the original tank for diesel and the new tank for the vegetable oil.

Figure 8.25
Large plastic tank being installed into a van. To finish the installation, the tank must be securely retained with some straps.

Figure 8.26
Small plastic fuel tank.

Putting together your own two-tank system

Buying a quality commercial two-tank kit is the easy option but it is possible to put together your own kit, and most people who do use a preassembled kit end up modifying or improving it. For the time-rich, money-poor, engineering-minded SVO enthusiast, assembling one's own kit is an exciting and challenging project.

You will need to put together a list of components. Beware, there are a lot of very substandard kits and bits of kits on the market; remember, generally you get what you pay for. Also, beware of DIY plans that may be nothing more than what you could find on the Internet yourself.

At the very least, you will need to gather together:

- A second fuel tank: bought, from an old boat or made-to-measure?
- A heat exchanger: bought, from junk yard or made yourself?
- A second filter: heated?
- Valves/solenoids: bought or from a junk yard? Trucks often have them.
- All the wiring, fuses, plumbing, hose, hose-clips, and so on, that you will need. Plastic pneumatic pipe works well for long runs, with little metal inserts to stop the pipe collapsing when hose clips are tightened. Use fuel-resistant rubber for short distances.

Always ensure all parts are resistant to diesel, biodiesel, and SVO.

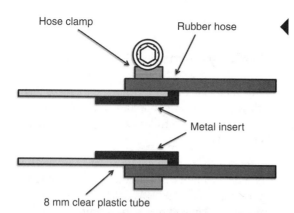

Hose clamp

Rubber hose

Metal insert

8 mm clear plastic tube

Figure 8.27
Diagram showing how to connect pneumatic pipe to hose.

Figure 8.28
Hose clips and pneumatic pipe inserts.

Figure 8.29
Pneumatic pipe run alongside existing fuel pipes on one of the authors' van.

Commercially available two-tank systems

There are dozens of commercially available two-tank systems on the market. Many of them are of excellent quality; however, some of them are not so good, and some of them are appalling – you get what you pay for. The German kits seem to lead the way, but there are also some excellent and popular UK and US products.

Most manufacturers sell a range of similar designs, which vary depending on your engine's size (larger engines will need larger heat exchangers and wider pipes), the amount of space you have available for your second tank, and whether your vehicle is 12 V or 24 V. Other two-tank kits are engine- or vehicle-specific, with components designed to fit and made-to-measure fuel tanks.

A typical good-quality two-tank kit will consist of a second fuel tank (which you will need to select depending on your vehicle and how much space is available), a second fuel filter, a heat exchanger to plumb into your engine's cooling system, solenoid-actuated valves to switch between tanks, and all the cabling, hoses, relays, and switches you will need. Some two-tank kits also come with special new glow plugs, temperature sensors, and "computers" to manage the system. All should come with full and easy-to-understand fitting instructions and operator's instructions.

In terms of vegetable oil conversion kits, it really is a case of "The Good, The Bad, and The Ugly." There are some fabulously good kits available as well as some astoundingly bad ones. The following represent a few, but certainly not all, of the better conversion kits on the market; just remember that you tend to get what you pay for:

Elsbett

High-quality German kits designed to fit specific vehicles. There is a list of their international resellers on their website.
www.elsbett.com

ATG/Diesel-Therm

ATG's Vegetable Oil-Kit, utilizes their Diesel-Therm electric heater, along with a heat exchanger, to ensure SVO is up to temperature. There is a list of their international resellers on their website.
www.diesel-therm.com/vegetable-oil-kit.htm

Smartveg

Smartveg are U.K.-based, they produce two-tank SVO systems with computer-controlled fuel switching.
www.smartveg.com

Golden Fuel Systems (formally Greasel)

Golden Fuel Systems (not to be confused with Golden Fuels in the U.K.) are U.S.-based and produce coolant-heated two-tank SVO kits.
www.goldenfuelsystems.com

Frybrid

Frybrid is a U.S.-based maker of two-tank SVO systems with computer-controlled fuel switching.
www.frybrid.com

One-tank systems

In one-tank or single-tank systems the engine is started and stopped on vegetable oil alone. There is no need to switch between fuels, no waiting for the engine to heat up, and no need to remember to switch back again before you arrive; you just fill the existing tank with vegetable oil (or indeed mineral diesel or biodiesel or any combination of them) and drive just like any other vehicle.

In addition to modifications to the fuel system similar to the two-tank method, this solution relies on the modification of the injection system to better suit the characteristics of vegetable oil. Typically, the injectors spray pattern are modified and "cracking" pressure is increased (typically by 5–20 bar) and larger, hotter glow plugs are installed which stay on for a while after the engine has started.

Figure 8.30
Two Elsbett converted VW vans, and Tinker, one of the authors' dog, in Wales.

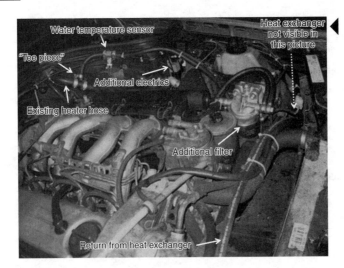

Figure 8.31
Elsbett converted Mercedes engine.

Advantages and disadvantages of single-tank conversions

Only specific engines can be converted using a one-tank kit and "DIY kits" are not are really an option. The conversion is a little more expensive and a little more complex than a two-tank conversion, as you have to make minor alterations to the engine as well as the fuel system. Many kits also require the injectors to be sent off to be altered, but this is a very quick process and the kit should include a prepaid envelope for this. Once installed, conversion is not so easily reversible and the kit cannot be transferred to another vehicle unless it is of the same type.

The advantages are clear: only one tank to worry about, no waiting for the engine to heat up, no switching over and switching back, fill up with SVO or diesel or biodiesel or any combination of them, all in the one tank.

One-tank DIY

Attempting a do-it-yourself single-tank conversion should only be done with extreme caution. It can be done, and indeed has been done successfully, but generally only by very keen, very competent enthusiasts.

Unlike twin-tank conversions, when done properly single-tank conversions involve modifications not only to the fuel system but also to the engine's fuel delivery system. The injectors themselves are modified and the glow plugs upgraded. In kits such as the Elsbett kit, the injector nozzles are specifically made; they cannot be easily made by the average SVO enthusiast.

List of things to think about:

- Increase the diameter of the fuel lines
- Heat exchanger
- Electrical fuel heater
- New injector tips
- Increase injector's crack pressure
- Change glow plugs for longer hotter ones
- Alter glow plug's relay so they stay on after start-up, typically 3 minutes

If in doubt, consider a one-tank kit or the other options in this book.

Commercially available one-tank systems

The one-tank conversion kit market is dominated by Elsbett, and less so by one or two less well-known companies. There are also a number of Elsbett imitators, too, who are selling rather substandard copies of the Elsbett kits, some of which leave out vital bits such as the modified injectors.

Elsbett

Elsbett has more than 30 years at the forefront of using vegetable oil fuel in diesel engines. In 1979 Ludwig Elsbett developed a three-cylinder air/oil-cooled diesel engine specifically for vegetable oil; in many ways it showed the way for all modern DI engines. Elsbett now makes a range of single-tank and twin-tank conversion kits for specific engines and vehicles; check their website to see if yours is one.

If an auxiliary heating system or night heater is desired, then a small secondary tank should be fitted for diesel, as these heaters do not run on vegetable oil.

www.elsbett.com

VWP, Vereinigte Werkstätten für Pflanzenöltechnologie

Combined workshops for vegetable oil technology: provides advanced single-tank SVO systems.

www.vwp-europe.com

WOLF Pflanzenöltechnik

WOLF vegetable oil technology: provides advanced single-tank SVO systems.

www.wolf-pflanzenoel-technik.de

Other single-tank kits

There are many other single-tank conversion kit manufacturers, worldwide. As ever, do your homework: Don't believe everything you are told!

Single-tank conversion installation example

Fitting the Elsbett one-tank kit into a VW Transporter, type 4/T4 with a 2.5 Tdi engine

This Elsbett single-tank conversion kit uses the heat from the engine's coolant water to heat the SVO up to a temperature where it is of a similar viscosity to mineral diesel. It also uses

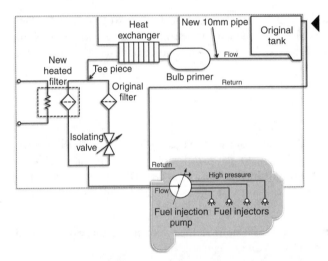

Figure 8.32
Schematic of the modified fuel system.

Figure 8.33
The engine on a WV Type 4 is accessible as the radiator can easily be swung out of the way. (Courtesy of Robert Starbuck.)

wider fuel pipes, modified injectors, an additional heated fuel filter, and upgraded glow plugs that are kept energized for longer after the engine is started.

Accessing the engine on these VWs is a breeze, as the radiator can be swung out of the way. This done, the engine is checked for problems, cleaned, the battery disconnected, and the water drained.

Figure 8.34
Close-up of fuel injection system. (Courtesy of Robert Starbuck.)

Figure 8.35
Close-up of fuel injection system with high-pressure pipes removed. Note the tape preventing dirt getting into the injection pump. (Courtesy of Robert Starbuck.)

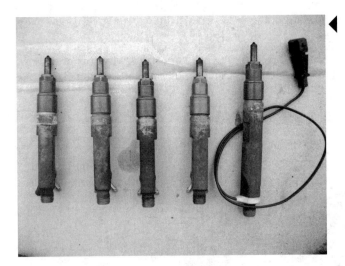

Figure 8.36
The five injectors removed from the engine. (Courtesy of Robert Starbuck.)

The injectors are removed from the engine and packaged up in a prepaid envelope and sent to Elsbett for modification. It is a good idea to do this first, as, while they are quick in returning them, it will nevertheless take some time. While they are doing this, it is important to keep the injection system clean: Cover exposed parts with tape, and block the holes where the injectors came out to ensure you don't drop anything in there. When you are ready to put the injectors back into the engine, be sure to use the new seating washers supplied by Elsbett, not the old ones or new standard ones.

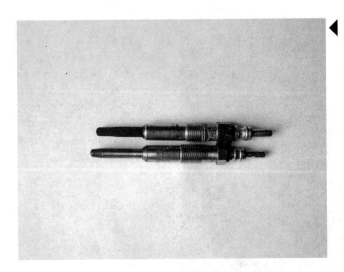

Figure 8.37
Old glow plug (top) and new glow plug (bottom). (Courtesy of Robert Starbuck.)

Replace the old glow plugs with the new ones supplied. They are slightly longer and heavier duty. They will be held on for up to several minutes, rather than just a few seconds, in the converted engine.

Figure 8.38
Heat exchanger and auxiliary fuel filter are mounted in engine bay. (Courtesy of Robert Starbuck.)

Choose carefully where you want to mount the heat exchanger and the second fuel filter. You may need to make up some new brackets for this. They must be mounted away from heat sources, such as the exhaust system, and away from places they may get wet or damaged. It is important to mount the heat exchanger the correct way up, so that any air in it is not trapped. Also consider how easily it can be accessed in the future, as you will need to change the filter at some point.

Figure 8.39
Connecting the new fuel pipe to the existing fuel tank. (Courtesy of Robert Starbuck.)

The fuel pipes in this conversion are swapped from the standard 8 mm to a larger 10 mm; this has to be done all the way from the engine back to the fuel tank.

Figure 8.40
Plumbing finished on the injection pump. (Courtesy of Robert Starbuck.)

Plumb the new components onto the fuel pump: There are now two fuel feeds, one from each filter. These are attached together, one on top of the other, so that they both now feed the pump. Compare Figures 8.34 and 8.40. In Figure 8.40 notice how the twin fuel supply pipes (bottom right) and the fuel return pipes – one from the injectors and one from the fuel pump – differ from Figure 8.34.

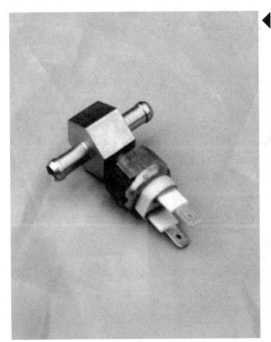

Figure 8.41
Close-up of coolant water temperature sensor.

Figure 8.42
Coolant water temperature sensor in heat exchanger's hose (near center). Also visible are the T-pieces into the van's cabin heater (near top right and near bottom left). (Courtesy of Robert Starbuck.)

Figure 8.43
Heat exchanger and second filter installed and plumbed in. Also visible is the filter isolator valve. (Courtesy of Robert Starbuck.)

◀ Figure 8.44
Existing fuel filter plumbed back in. (Courtesy of Robert Starbuck.)

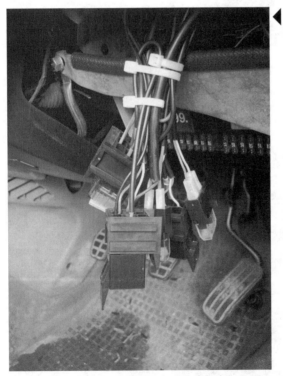

◀ Figure 8.45
Wiring loom exposed under vehicle's dashboard. (Courtesy of Robert Starbuck.)

The heat exchanger is plumbed into the cabin heater circuit in parallel using T-pieces along with a temperature sensor.

The second fuel filter is run in parallel with the existing fuel filter and an isolating valve is added to the existing filter; this acts as a backup filter. The other filter is also equipped with an electrical heating jacket to ensure the fuel in it is sufficiently liquid to pass through it.

The kit also comes with a priming bulb, which makes bleeding the system very much simpler.

There are a number of electrical components and relays in the kit that need to be carefully wired into the existing wiring loom. These sense the engine's temperature and control the glow plugs and the fuel filter heater jacket. A wiring diagram comes with the kit with standardized numbering (such as 15 = 12 V ignition, 30 = 12 V positive permanent, 31 = chassis, ground, and so on) but you will need to find an accurate wiring diagram for your vehicle yourself.

Chapter 9

Using WVO in your SVO engine

Introduction

Collecting waste or used oil, WVO or UCO (see Chapter 4) and using it directly in your SVO converted engine is a very attractive prospect; your fuel is practically free and you are finding a use for what is often a waste product, possibly destined for landfill or illegal dumping. However, as with many things in life that seem too good to be true, this could be too.

Elsbett, for example, is quite specific: You must only use vegetable oil that meets the required standard with their kits, which definitely excludes used oil. Many people ignore their advice and use cleaned-up WVO, but it is important to understand the issues and manage the risks.

You have to weigh up the pros and cons of running on used oil over new oil. Obviously, there is the price of the fuel to consider. Used oil is going to be very cheap when compared with new oil, but it is not going to be free, as you are going to have to invest in apparatus to clean it and it could reduce the life of your engine significantly. Consider the age of your engine. How much it will cost to replace? What is the cost of new vegetable oil and the cost and space implications of cleaning used oil? Ask yourself if you are a practical minded person who loves to tinker with his or her older car? Or are you simply trying to save money or green your new and expensive SUV?

Using poor-quality WVO or not filtering and dewatering it is very likely to be very much more expensive in the long run than just doing it right in the first place.

At the same time, new oil, SVO, comes with a lot of social and environmental baggage; it is not as green or as ethical as many people would like you to believe (see Chapter 14 for more on this.)

Processing waste oil

Whether your intention is to make biodiesel or just use your WVO straight, you will still need to clean your oil. In the case of making biodiesel, other than prefiltering most of the filtration is normally done after the reaction. If you intend to use WVO as a fuel in an SVO-converted vehicle, you will need to clean it up first. Various levels of filtration, dewatering, and testing for acidity need to be done to convert dirty waste oil into usable fuel, and even this is not necessarily great-quality fuel.

Test first, WVO or biodiesel?

The quality and condition of the used oil you collect is much more important when you are thinking about using it in an SVO system than if you're going to convert it into biodiesel. The differences between new food-grade oil and heavily used oil can be large.

All but the least-used waste oil will contain significant concentrations of free fatty acids (FFAs), acidic components that your engine is not designed to resist. Slowly, over time, these acids may corrode internal components of your fuel system and engine, possibly causing them to eventually fail.

As well as FFAs, there will probably be significant amounts of water in your WVO that will not boil off easily and, if there is enough of it, will inhibit combustion and damage your engine. There is a relationship between FFA content and water content; if you have a lot of FFAs, you probably have a lot of water, and vice versa.

Test the WVO you have collected (or even better test it before you collect it – that way you can leave it alone) for FFAs using the titration test in Chapter 4: the lower the titration value the better-quality fuel it is.

There is no clear rule as to what is a too high titration value, but a rule-of-thumb is that it may be best to avoid WVO titrating at more than 3 or 4 in an SVO system. However, just how acidic you feel is OK and is ultimately up to you. If the oil titrates higher than you would like, don't despair. It does not make it useless: maybe you could always process it into biodiesel instead (see Chapter 6).

What is filtering?

To filter means to separate two or more things. The sense that we mean it here is to take dirty, used cooking oil and to separate the solid particles, which we don't want, from the waste oil, which we do.

Unfiltered or badly filtered oil put directly into the fuel tank will quickly block the vehicle's own filter; we need to remove all the muck before we put it into the tank. Don't think of your engine's built-in filter as a fuel cleaner; think of it as the engine's last line of defense against the rubbish in the fuel.

If particles do get into the engine, they will cause damage. The larger particles may block fuel lines, whereas the smaller ones will block the tiny holes the fuel has to pass through, such as the injectors, and damage parts of the injector pump that require the fuel to lubricate them.

A micron, or micrometer (μm), is one-millionth of a meter, a very short distance indeed; e.g., human hair is roughly 50 microns across. Engines typically come with filters that remove particles down to 10 microns, often better, so we need to get the WVO at least this clean and preferably cleaner before we can even think about putting it into the tank.

> Some SVO kits come with worrisome high-gauge filters of more that 10 microns, some as high as 40 microns. Putting a high-value filter as your final filtration is not a good idea; engine manufacturers put filters of certain values on their engines for good reasons, and not following their lead will inevitably cause damage to your engine. If you find your filter is becoming blocked by your fuel, fitting a larger-gauge filter is not addressing the problem. Depending on what is causing the blockage, you may need to fit a filter heater or re-examine your WVO cleaning regimen or check your vehicle's tank and fuel lines for contamination.

Commercial biodiesel makers often heat the oil prior to filtering and use pumps for speed, whereas the small-scale operator will tend to prefer the slow methods and get gravity to do most of the work.

Prefiltering

First you will need to remove the big bits: the French fries, insects, and dead rodents. Any kind of mesh will do, from a special screen that fits on the end of the hose you suck the dirty WVO up with to pouring it through an old mesh satellite dish. There are also strainers available at 200 and 100 microns that fit the top of 20-liter buckets and 205-liter drums.

Settling

If you have the time and the space, leaving your collected oil to settle is a very good idea: Simply let the oil stand, undisturbed, for as long as possible. After a few days, but preferably weeks, the heavy particles will tend to sink to the bottom of the container and you can siphon or pump the cleaner oil from the top of the container, ready for filtering properly, and drain the dirtier stuff from the bottom.

Standpipe tanks are ideal for this: See Chapter 7 for how to make one.

This is also a very effective way of getting rainwater from the oil but does not get all of the water out.

Bag filters

Bag filters are a popular, simple, and cheap solution. A bag filter is a sock-shaped bag about 70-cm long with a 10-cm wide opening at one end, and can cost as little as $5. Bag filters come in a variety of filtration sizes.

◀ Figure 9.1
Bag or sock filters. (Courtesy of biodieselfilters.co.uk.)

Bag filters can be mounted in specially made enclosures, but these are rather expensive, so most people either make their own enclosure from PVC pipe fittings or even simply cut some holes in the top of a 205-liter oil drum and pour the oil in.

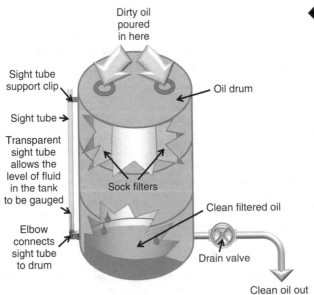

◀ Figure 9.2
Sock filters in a converted 205-liter barrel: labor intensive, but simple and cheap.

In some cases bag filtering is the only cleaning done to the oil, but further processing is preferable, especially since your oil will almost certainly have water in it.

Cartage filters

Cartage filters are a popular option, often in combination with a pump, but sometimes in gravity-fed systems. They come in a vast number of shapes and sizes, some with paper elements like the ones you get in cars, but usually bigger, and some with polypropylene or nylon elements.

Figure 9.3
Polypropylene filter housings. (Courtesy of biodieselfilters. co.uk.)

It is a good idea to use several filters in series, starting with the coarsest and going down to progressively finer filters. This stops the finer filters getting blocked with large bits of debris. Start with maybe a 100 or 200 micron, then about a 50 micron, and then a 10 micron and possibly finer.

In Figures 9.5 and 9.6 pressure gauges (one on each filter) indicate the pressure in each housing as the oil is pushed through by the electric pump. They enable the operator to determine which filter needs to be changed as the pressure in the housing will rise as it becomes blocked. The design shown would be greatly improved with the addition of valves before, between, and after the filters to enable the elements to be removed and replaced as they became blocked without the need to drain down the entire system.

Figure 9.4
One of the authors' early experiments in gravity-fed filtration. On the left is a water-cooler bottle feeding into a polypropylene cartage filter and on the right is a plastic carboy feeding into an old engine's fuel filter.

 www.biodieselfilters.co.uk stock a wide variety of filters.

Dewatering your WVO

There are two principal ways that water can get into your WVO: from cooking and from the rain. The former is inevitable, whereas the latter is avoidable. Water is bad. If you are making biodiesel it will upset your reaction and if you are using the oil directly in an SVO system it will damage your engine and it won't help the fuel burn at all.

As described in the Chapter 5, vegetable oil is made of *triglycerides* or fats, a subgroup of molecules called *lipids*. A triglyceride can be thought of as a glycerine molecule with three arms: These arms are *fatty acids*.

Figure 9.5
One of the authors' more advanced WVO setup, with settling tank, pump (bottom left), and three levels of filtration.

Using the oil for cooking causes damage to the triglyceride molecules: Their arms break off; they become *free fatty acids* (FFAs). If one arm breaks off the triglyceride, you have a two-armed triglyceride, now called a *diglyceride,* and an FFA molecule. If two arms get broken off the triglyceride, you have a one-armed triglyceride, a *monoglyceride* (more correctly a *monoacylglycerol*), and two FFA molecules. Break all three arms off and you have a glycerine molecule and three loose FFAs.

The bond between the FFA and the glyceride is called an *ester.* When making biodiesel the three fatty acid legs of the triglyceride are detached from the glycerol molecule and attached to the methanol molecule, making a *fatty acid methyl ester* (FAME) or biodiesel, and glycerol as the by-product.

Monoglycerides, diglycerides, and FFAs are all *amphipathic lipids*; like triglycerides they are mostly *hydrophobic* or water-repellent, but unlike triglycerides they have a region that is *polar, hydrophilic,* or water-loving.

Figure 9.6
Close-up of filters and their pressure gauges.

Monoglycerides and diglycerides are used in the food industry as emulsifiers: i.e., they are used to make fatty foods mix with watery foods. such as mayonnaise, where a watery food, in this case egg, is mixed with an oily food, oil.

So what does all this mean?

While water *mixed* in with new oil will separate easily because its triglyceride molecules are hydrophobic or water-repellent, used oil, with its monoglycerides and diglycerides, will not give up its water so readily because some of its molecules are partly hydrophilic or water-loving.

As we have discussed in the biodiesel chapters, well-reacted biodiesel will not have many FFAs, monoglycerides, and diglycerides, only fatty acid methyl esters, and it gives up any water in it quite readily. However, your WVO is not biodiesel. It may want to hang on to its water: They may be hard to separate.

When using WVO in an SVO system, the FFAs, monoglycerides, and diglycerides in the WVO may also come with significant amounts of water. In addition, the FFAs are acids (hence the name), and these may slowly damage your engine and fuel system.

Testing for FFAs

Use a titration test (see BoxOut in Chapter 4) to determine the FFA content, and hence estimate the monoglyceride and diglyceride content.

Testing for water

This section is repeated in Chapter 4.

Testing for water in your WVO is simple, if rather crude. Take a small cupful of your oil in a pan and gently heat it to around 100°C (212°F). If there is bubbling or spitting, you have water present. How much is usually a matter of guesswork and experience.

Quantitative test for water

You know you have water present, but how much? Calculating how much water is in your WVO is not difficult. Take a representative sample from your WVO by first ensuring it is well mixed. Take a few hundred milliliters, measured quite carefully, and weigh it accurately. Put it into a pan and heat it gently; let the water in it boil, but not so much as it spits out of the pan. Careful here: Hot oil is the number one cause of household fires and can give you nasty burns. When it has stopped spitting, turn off the heat, allow it to cool, and weigh it again. The difference between the two numbers is how much water you had, 1 ml of water weighs exactly 1 g, so if you know how many milliliters of wet oil you started with, you can calculate how wet the oil was as a percentage.

However, a quantitative test is of limited use because, whether you're making biodiesel or using the WVO straight, anything more than the tiniest amount of water is not acceptable.

Dewatering

By prefiltering first

A lot of the water in your WVO is attached to the particles of food in it. Simply removing the particles, by mecanical filtration and settling, will remove a lot of the water too. If you stand the oil in settling tanks, as described earlier, along with the bigger solid bits of rubbish in the WVO, some of the water will sink to the bottom of the settling tank and can be drained off.

If you have water in your WVO, use this method first and then retest for water, as described above, to gauge how much, if any, you have left to remove.

Dewatering your WVO further

To dewater your oil further, you can heat it all to a temperature higher than the boiling point of water (100°C, 212°F) and hold it there until it all the water has boiled off. However, for any significant quantity of oil, this is slow, requires a very significant amount of energy and is somewhat dangerous and a fire hazard.

A safer and less energy-hungry method is to heat the oil to around 60°C (140°F), put it in an insulated container, such as a hot water tank or a 205-liter drum surrounded by insulation, and leave it overnight. At this temperature, the water is significantly denser than the oil and will sink to the bottom more readily than with just a cold settling. Do not be tempted to keep the oil warm overnight by leaving the heater on, as this will cause the oil to circulate; the oil needs to be free to stratify. In the morning, drain off the bottom 15% or so and do not use it; retest the top portion for water, using the test above, as it may well still have significant amounts of water in it; if it does, dewater again, and consider getting some better WVO next time.

For additional ideas on dewatering, see Graham Laming's biodiesel processor in Chapter 7.

WVO in the cold weather

SVO is less winter-hardy than biodiesel, which itself isn't very winter-hardy. Vegetable oils have higher cloud points at which they start to gel (turn solid) than biodiesel made from the same oil, and this is even truer for used oils. As the weather gets colder, your WVO fuel will begin to freeze. Initially, it will become more viscous, and then little crystals of frozen fuel will appear. As it gets colder still, all the WVO will freeze to a thick jellylike state. At what temperature this happens is entirely dependent on your oil.

As with SVO, the addition of about 10% diesel to your WVO will probably do the trick, but do it before the weather gets cold, not after, or you will have a tank of usable fuel while the engine's pipes and filters will remain blocked. See Chapter 8 for more ideas such as heated fuel tanks, pipes, and filters.

Summary

To summarize this chapter, using WVO in an SVO system is a great idea because you are turning waste into a cheap fuel; once you are set up to process it, it is a very cheap fuel. What is more, you don't have to worry about the environmental and ethical concerns over using new oil as fuel.

However, there is a significant amount of work involved in turning WVO into fuel suitable for an SVO converted engine; it must be thoroughly cleaned, filtered to at least 10 microns, and properly dewatered.

Even then, it may not be such great fuel. It may be sufficiently acidic to damage your engine. You need to test its FFA content: If it titrates higher than about 3 or 4, then you maybe should be thinking about turning it into biodiesel instead.

Chapter 10

Other options and solutions

This is the weird chapter in this book – included as an attempt to be complete but not necessarily to recommend. As we have seen, the basic difficulty when getting a diesel engine to happily run on a vegetable oil is viscosity. We need to reduce the fuel's viscosity to something similar to the viscosity of mineral diesel fuel; with biodiesel, we chemically modify the fuel to be more compatible with the engine; with SVO, we modify the engine to be more compatible with the fuel.

However, this is not the end of the story; there are a few diesel engines which are remarkably tolerant of a wide range of fuels and most diesel engines (especially the older ones, less so the most modern ones) are at least a little tolerant of fuels that they were not entirely designed for.

Many people claim to have found a variety of different "wonder solutions" to making different fuels work in a diesel engine. Some of these solutions will work, but it is more a credit to the fuel-tolerant nature of the diesel engine than any particular technical mastery on the part of the inventors. Many will work for a while but could ultimately result in long-term damage to your engine.

It is important to bear in mind your motivation for running your engine on alternative fuels. If you are doing it in the pursuit of being "more sustainable," think carefully about the solution that you adopt. Many of these additives are derived from petrochemicals and, furthermore, may cause unwanted emissions from the exhaust of your vehicle.

Cetane number

The cetane number measures how well a fuel will autoignite under compression – something you want in a diesel fuel especially in the winter. The octane number of a fuel is the measure of how much it resists autoignition under compression. It is the opposite of the cetane number; this is something you want in a gasoline engine but not in a diesel one and not vice versa.

About 40–50 is a typical cetane number for normal diesel fuel.

Diesel, kerosene, and heating oil

Some people run their diesel engines on fractions of crude oil that are similar to diesel oil, such as heating oil, kerosene (paraffin), or blends of these and other oils, with varying rates of success.

In the hierarchy of fractions of crude oil, diesel is somewhere in the middle. At one end you have very low-viscosity fuels such as aviation fuels and gasoline and at the other end you have the heavy viscous fractions such as waxes and asphalt. In the middle you have diesel fuel, with slightly lighter kerosene on one side, which oil lamps burn, and the heavier industrial (bunker) fuel oil, which big ships run on.

Diesel fuel itself is refined into a number of finer grades: number one diesel, number two diesel, and number four diesel; number three diesel is no longer made. Number one diesel is the least viscous, number four the most viscous, but there is not a great deal of difference between them.

You will find number 4 diesel in lower-speed diesel engines such as generators and railway train locomotives, where the engine's speed is fairly constant; numbers 1 and 2 are common as "normal" diesel for road use, number 1 being premium diesel, with better winter characteristics, number 2 diesel being not so good in the winter but better at lubricating the engine. In addition, the diesel that you buy at the filling station will have any number of additional additives and improvers:

No. 1 diesel = Cetane number of about 45–50

No. 2 diesel = Cetane number of about 40–45

Kerosene is slightly thinner than diesel and, while it combusts similarly to diesel, it is too poor at lubricating for use as fuel in a diesel engine. Heating oil is probably the closest to number 2 diesel in both ignition and lubricating qualities and is often used as a diesel substitute; however, while it does seem to work for some people, it is not diesel and your engine was not designed to run on it.

Some people do use kerosene in blends with vegetable oils, and often other ingredients such as mineral spirits and even gasoline, as a substitute for diesel fuel. You can see why too: Kerosene is too thin for a diesel engine and vegetable oil too thick, but perhaps a blend of the two would make a good substitute? Generally, it does work, at least for a while, but there has been very little research done beyond anecdotal reports into what ratios of oil make a good blend or, in fact, whether it makes a good fuel at all.

Not only is running on kerosene or heating oil potentially risky to your engine, it is often illegal. Depending on the country you are in, heating oil and kerosene attract lower rates of tax or duty than road fuels and they are often marked with a color indicator to stop you using the wrong one. For example, in the UK, industrial diesel not intended for road use (red diesel, used in farm machinery and generators) is taxed at a lower rate from road diesel (DERV, diesel engined road vehicle); in just about every other respect the two are identical and clearly there is a temptation to run your road vechicle on the cheaper, offroad, fuel. The two are distinguished from each other by dyeing the red-diesel red, hence the name; use the wrong one and when your tank is "dipped" by officials they can see what you have been up to and the penalties are heavy. The same often applies to heating oils and kerosene; they often have color, indicators added and show the tax man you have been evading tax.

Solvent thinning and blending

There are many thousands of advocates across the world of blending vegetable oil with any number of solvents with the intention of lowering the viscosity, and altering the cetane number.

Search the Internet for recipes but be very cautious and don't believe everything you are told.

The option is very attractive when compared with biodiesel or SVO conversions, with no complicated chemistry or processors or by-products or expensive engine conversion; all you have to do is obtain some vegetable oil and mix it with some solvents and you have a biofuel diesel substitute. However, there are so many variations and recipes and nonsense misinformation out there that it should be approached with extreme caution; please treat these fuels as, at best, experimental.

White spirit

White spirit is an example of a thinning agent used with vegetable oil to make a diesel substitute. It is used as another petroleum distillate. If you remember back to the diagram of the fractional distillation column in Chapter 1, you will realize that lighter fuels are tapped off high in the column. The hydrocarbons in white spirit generally have chains in the range of $C_7 - C_{12}$. In addition to straight hydrocarbons, which we have discussed in this book, white spirit also contains alicyclic components, where the carbon atoms form a ring.

White spirit is available in several types. The most basic type contains a fair amount of sulfur, which gives it a low flash point. This is known as "type 1" or T1. Type 2 white spirit has some of the more volatile solvents removed, whereas type 3 white spirit, which has the highest flash point, is hydrogenated (explained in Chapter 5).

If your interest in finding substitutes for fossil diesel is to be green then don't forget that kerosene and heating oil are just as much fossil fuels as diesel and gasoline, and that many gasoline solvents are derived from petroleum; e.g., white spirit (mineral spirits) and terps substitute (mineral turpentine) are both made from petroleum.

Burning gasoline, diesel, or vegetable oil in your engine is not excatly clean, but many of the solvents suggested as ingredients for blended biofuels may make some additional, very nasty or toxic, compounds.

The Infopop forum has a whole section on experimental thinning of vegetable oil with solvents as alternatives to biodiesel and SVO engine conversions: biodiesel.infopop.cc

Open source, not secret sauce

The biodiesel community, especially with the advent of the Internet, is almost entirely an open-source community information that is freely available, noncommercial, and in the public domain.

Whereas there is certainly much debate about making or using biodiesel or converting your engine to SVO, none of it is a secret being withheld from anyone, anywhere.

Indeed, there is nothing in this book that you can't find for free on the Internet. However, we think this book is more concise, easier to read, technically sound, clearly illustrated, and, most importantly, more objective and more accurate than most of what we've come across online. That said, there are a number of companies who will, for a price, sell you their "secret recipe" for their biofuel, which is usually not biodiesel or SVO but their own secret blend. There are also plenty of secret additives on sale that will, allegedly, "improve your fuel" or turn your vegetable oil into biodiesel with no "complicated chemistry."

2-Ethylhexyl nitrate

2-Ethylhexyl nitrate (2EHN) is a common additive to mineral diesel as a cetane improver. Some people add it to biodiesel and vegetable oil and claim it improves things, especially cold starts. It is also added by people who make vegetable oil blends for diesel engines with gasoline or alcohol, which reduce the cetane number, as a "cetane restorer," and it is almost certainly helpful.

That is too good to be true!

When something seems too good to be true, then chances are it is sadly just that: It is not true. According to numerous sites on the Internet, you can buy all sorts of miracle products or plans to improve your mileage and reduce your fuel bills. Don't believe everything you are told. Question everything and don't take anyone's word for anything.

Anecdotal evidence is not good evidence

In the same way that we all know smoking causes cancer and yet we all also know of a smoker who lived until he was 110 years old, you will come across people who will claim to have run their unconverted car on used oil with no problems – just because they had no trouble does not mean you won't.

Extraordinary claims require extraordinary proof

If I told you that I drink one cup of my special tea each day you would probably believe me; after all it is not unlikely and even if I were lying it does not mean that tea and drinking don't exist. However, if I told you that drinking one cup of my special tea each day meant I would never suffer from heart problems you would be sensible to want more proof than to take my word for it, and before you invested your savings into my tea-drinking regime you would want to see independent double-blind peer-reviewed medical studies backing up my claims. Nevertheless, every day people are duped by charlatans and con artists into parting with their cash in exchange for some ludicrous invention or theory that has only the flimsiest evidence to back it up. In short, don't believe everything you are told and demand to see real evidence, not just anecdotal; remember, the more outlandish the claims the better the evidence needs to be.

Magnets

There are special magnets made which attach around your fuel lines and align the fuel molecules, resulting in better combustion, better mileage, and lower fuel bills. Sounds like nonsense to me, but thousands of people buy them every year; some even claim they work! The suppliers offer vaguely convincing sounding, pseudoscientific language to back up their claims but the words are usually meaningless. Think about it: If these things work, in these days of demand for high-mileage, fuel-efficient, and greener cars, they would be fitted as standard. Indeed, politicians would pass laws making them compulsory.

HHO and water-powered cars

There are machines available, often called "HHO Technology," that convert water into its component parts (hydrogen and oxygen) and introduce these into your engine's air intake and drastically reduce the frequency that you need to buy fuel because now your car runs, at least partly, on water. Does this sound too good to be true to you? Of course, but people still fall for it all the time! It takes a lot of energy to separate water into H and O, a lot more energy than there is available in the H and O when it is burned. And the energy it takes to do the separating is coming from somewhere; it is coming from your engine and therefore from your fuel. Indeed, if this "technology" did work it would be a perpetual motion machine, your engine would be starting its cycle with water (as fuel) and ending it with water (as exhaust) while extracting energy from the water, and this violates the first law of thermodynamics.

HHO, and several other fuel-saving scams, are neatly dealt with at New Zealand's Aardvark Daily where there is an unclaimed one-million dollar prize available for anyone who can show their HHO system saves them fuel:

www.aardvark.co.nz/fuelsaverscams.shtml

www.aardvark.co.nz/hho.shtml

Acetone

Acetone (known to your mum and sister as nail varnish remover) is added by some to either gasoline or diesel at about 1 part acetone to 500 parts fuel. Proponents talk of big oil company cover-ups and a dramatic improvement in fuel economy and engine life; however, the authors are unaware of any evidence to support this and, in fact, a little evidence to suggest that it has no measurable effect on performance and degrades engine parts.

Adding engine oil or automatic transmission fluid to diesel

The authors have seen many references to adding engine oil or automatic transmission fuel (ATF) to your diesel to increase its engine lubricating qualities; however, we can find no evidence whatever that it has any beneficial effect.

The great gas conspiracy

The great gas conspiracy is a theory that the big oil companies and the big car makers have a deal whereby, despite having technology to the contrary, they will continue to sell us fuel-inefficient vehicles and split the profits between them. We've never seen any evidence to support this! The fact is, vehicle manufacturers make big ol' trucks, because they are cheap to make and profitable to sell and people buy them, and the oil barons aren't going to do much to stop them; however, that doesn't mean they are in cahoots. Independently, though, the oil companies and the gas companies aren't doing a lot to change the status quo. However, with growing consumer awareness of issues of sustainability, and in the knowledge that oil isn't going to last forever, this is more a result of ensuring continuity of business (and hence profits) than any real social goal. It is always the early adopters of a technology who innovate and drive things forward – waiting for the status quo won't encourage change, but being part of the change will.

Pogue's carburetor

Between 1928 and 1935 Charles Pogue applied for several patents for his carburetor that supposedly completely vaporized the fuel before introducing it to the cylinders, which would supposedly enable a great deal more energy to be extracted from the fuel. There were several reports at the time of people driving seemingly miraculous distances on small amounts of fuel, including a 1936 issue of a Canadian magazine that claimed that a trip of about 1800 miles was completed on 15 gallons of fuel using a Pogue carburetor. However, when credible people, scientists and engineers and the like, tried to see one of these carburetors or measure its performance they were not granted access; this nonsense went on for a few years until some articles were published rubbishing Pogue's claims. Basically, he was told to produce the carburetor or shut up, and he shut up. Lots more on this bizarre story can be found on the Internet: Both arguments claiming that the carburetor never existed and arguments claiming that the fact the world has never seen one is proof that it has been covered up.

Not a diesel engine?

Then why are you reading this book? Petrol or gas engines have several options to petroleum such as alcohol blends, and can be converted to run on compressed natural gas (CNG), liquefied petroleum gas (LPG), and so on. Then there are also electric and electric hybrids, hydrogen, wood gas, propane, methane, and fuel cells. We're not going to cover these technologies comprehensively – there are other books on these topics, but we'll at least indulge you by giving you an introduction to these technologies, so you know what they are about when you come across them.

Wood gas

Those looking for a very quirky project should take a look at wood power. A machine called a wood gasifier can be built, and people do, to turn wood or other biomass into a gas that can be burned in a fairly standard petrol engine. Just fill up on logs and off you go! Wood gasification works by heating biomass (wood) in a sealed container in the absence of oxygen.

This heat source can be provided by burning more wood. Wood contains lots of chemicals in the resins and sap and cell structures of the wood fiber. The process is effectively "making charcoal"; but in the process of doing this, all the volatile chemicals in the wood are driven off and can be collected and burned. Wood gas is predominantly a combination of hydrogen and carbon monoxide. This probably isn't a technology that is going to change the world – the world has moved on from this technology – so bear in mind that while technically possible, it is quite an impractical solution and definitely falls into the category of "esoteric."

Bioethanol

In a similar manner to growing crops, to extract oil, to run in a compression-ignition engine, there are alternatives for spark-ignition engines too. Bioethanol represents one alternative for the many gasoline engines on our roads. Crops are grown, which can then be processed to produce alcohol. These crops are rich in sugar content: e. g., sugar beets. Brazil has been doing this for more than 30 years, growing sugar cane and turning it into ethanol for use as a road fuel. The rest of the word is catching onto this idea and, good or bad, a lot of gas sold in the UK and the USA is partly ethanol from biomass and it is increasing rapidly. As we explore in Chapter 14, ethanol has acquired a somewhat dubious reputation. Government subsidies for farmers and the rapid price increases of bioethanol feedstock have made ethanol somewhat a free-for-all for producers wanting to make a quick buck, but the sustainability of "ethanol production" remains very much in question.

Figure 10.1
Saab BioPower 9-3.

Figure 10.2
Ford Focus Bio-Ethanol.

Liquefied petroleum gas (LPG or Autogas)

Liquefied petroleum gas is some of the lighter fractions that are produced when crude oil is extracted from the ground. Usually consisting of a mixture of propane and/or butane, LPG is a gas at atmospheric pressure; therefore, it is stored in strong steel containers. If we look at the chemical model of propane, way back in Chapter 5, Figures 5.5 and 5.6 (butane is in the same series but with an additional carbon), and compare it to say "octane" which has eight carbon atoms, we can see that propane is a relatively "hydrogen-rich" fuel. When we think about all the bonds that will break, releasing energy, contrasted to the amount of carbon (which will then react with the oxygen to form carbon dioxide), we can see clearly that propane contains less carbon for the same amount of energy output. However, it is still important to note that burning propane does still, at the end of the day, result in carbon emissions. However, propane is a very clean burning fuel, free from sulfur or other nasties such as tetraethyl lead (the lead in leaded fuel). It also has the advantage of being nontoxic and noncorrosive.

LPG has a very high octane number of 108 (compared to 95/97 for regular gasoline/petrol) and can run in an ordinary spark-ignition internal combustion engine, with some modification of the fueling system. Cars can be bought that run on LPG "out of the box" or conversions can be done to existing vehicles to "retrofit" them with an additional LPG tank. This takes up boot space, however, and gives the car the flexibility to run on dual fuels.

Because LPG is gasified by the time it reaches the engine, a special manifold must be installed which gives the choice between a standard carburetor or fuel injection for running on gasoline/petrol or a "vaporizer/regulator" for running on LPG. In the same way that we use the engine coolant in an SVO conversion to heat the vegetable oil gently to reduce its viscosity, the vaporizer/regulator heats the LPG gently to allow it to turn from a liquid into a gas. If your car is fuel injected, a device called an "emulator" plugs into the engine control unit – it "fakes" the signals that the car would normally receive from the fuel injectors when the vaporizer is in operation – because the injectors are redundant, but the ECU expects pulses from them; this stops your "check engine" light from coming on.

Figure 10.3
SMART LPG ForTwo.

Electric vehicles

Electric vehicles have come out of the days of the sluggish utility-style vehicles, and with modern battery technology, a new generation of electric vehicles is emerging that demonstrates that the technology is capable of generating very capable performance, with a range to suit most consumer's needs. In the past, one of the problems with electric vehicles has been the battery technology. Lead-acid batteries have been used in a lot of early electric vehicles; however, while it is a cheap technology that automobile manufacturers are familiar with (it's the same one used in the batteries of your car today), it isn't the solution required for vehicles with a performance to meet most consumer's needs. However, change is afoot! With the rapid development of battery storage technology, in order to support our ever more demanding compact electronic devices, battery technology has made rapid advances. Modern lithium-ion batteries are capable of storing a greater amount of electricity for a given weight than older battery technologies such as nickel cadmium, or nickel metal hydride technologies. There is also great potential for battery technology to make significant strides in the future, with nanotechnology opening up some interesting branches of materials science that could lead to ever greater storage densities.

One of the benefits of electric car technology is known as braking regeneration. When undergoing braking, the vehicle's brakes attempt to dissipate some of the energy of the car's forward momentum, slowing the car down. Standard brakes work by friction – a pair of pads contacts a drum or disk, providing high friction. The energy embodied in the vehicle's motion is thus converted to heat, which is why race car manufacturers spend a lot of money developing vented and drilled disc brakes in an effort to dissipate heat quicker and allow more braking to occur faster, more effectively.

Under urban driving conditions, your car is constantly stop–start–stop–start: Each time you need to brake gently, power is wasted to heat. Furthermore, having to rev your engine every minute or so as the traffic inches forward doesn't make the best use of your engine. Internal combustion engines are efficient within a relatively small "power band," where they produce optimum power and efficiency. Having to "blip" the throttle every now and again, and run the engine up to that power band where fuel is used most efficiently, does not make the best use of your fuel.

Figure 10.4
NICE Electric Car.

Hybrid vehicles

Hybrid cars take the concept that when an internal combustion engine (ICE) runs, it should do so at the best possible efficiency. It allows the engine to work within its optimum power band, by trying to "match" the load that the vehicle places on the motor more effectively, courtesy of an electric motor and battery pack. Say you are driving slowly: A hybrid car will take a small amount of power from its electric motor to drive the wheels. Electric motors provide large amounts of torque when they start, relatively efficiently, unlike a gasoline engine that has to be accelerated slightly for optimum pulling power. Now imagine you speed up. The vehicle's engine will start up and take up the load. If you're going slower than suits the engines optimum rpm, then the additional power will be used to charge the battery bank while the ICE is running.

Figure 10.5
Hybrid vehicle – Toyota Prius.

In addition, a hybrid drive train can do "clever things" like capturing the energy that would normally be wasted by braking and storing it in batteries for later use. The vehicle does this by using the electric motors to create a "mechanical load" (which produces electrical power) to slow the car, rather than conventional brakes, which create a mechanical load by creating friction between a brake pad or shoe and a brake disc or drum.

There are different types of hybrid drive trains: In some, the engine drives the wheels "directly" through the drive train with the electric motor as part of this drive train providing resistance (when generating) or assistance (when driving) as required. As the technology has developed, there are a greater number of different approaches to this problem, some using a combined "starter motor generator" in what could be considered a "mild" version of the hybrid.

Plug-in hybrids offer an interesting option that is being pursued vigorously by many automakers. In a plug-in hybrid, the vehicle connects to the grid at night to charge a larger set of batteries than would be found in a normal hybrid vehicle. The vehicle will run initially on electrical power - the idea that if you make short journeys, you might not need to start the gasoline engine at all! However, if you need to travel a longer distance, the ICE kicks in and the car operates like a "standard" hybrid.

Hydrogen as a future energy vector

Hydrogen has the potential to radically transform the way that we look at energy. At the moment we live in a "carbon-based" economy - our power comes largely from fossil fuels, which are based on a high carbon content - large amounts of coal, oil, and gas go into making the wheels of our industry and civilization turn. However, as we have discussed already, there are problems with carbon-based fuels, in that their emissions from burning could lead to a climate catastrophe. One solution advocated by many, is a transition to hydrogen as a fuel. The thing is, when you burn hydrogen, you don't create carbon dioxide – the only product is dihydrogen monoxide – H_2O, water. This isn't bad news – no nasty nitrogen oxides creating smog, sulphur to create acid rain, or carbon dioxide to cause the world to heat up – just plain old agua.

The problem comes, like with electricity above, with producing the energy to produce the energy carrier. See, this is the key distinction, unlike biodiesel, vegetable oil, mineral diesel or gasoline – hydrogen isn't a fuel. We can't go drilling for hydrogen and find a big pocket of it underground; we have to create it by breaking down other compounds. This isn't bad news, as there is so much hydrogen in the world, trapped in the seas as water ... all we need to do is liberate it. One way to do this is through using large amounts of electricity. Unfortunately, there has to be somewhere to get this electricity from in the first place. One solution is to produce clean, renewable energy from the sun, wind, waves, and tide, and produce benign electricity with minimal environmental impact; unfortunately, the other ways of generating electricity – fossil fuels and nuclear power, entail large amounts of carbon emissions and a legacy of toxic waste. So by using "dirty" power, all you are doing is extending the tailpipe of your vehicle to a centralized location.

There is another way to produce hydrogen, one that is championed by the oil industry; that is to produce hydrogen from a process called steam reformation. High-temperature steam and fossil fuels can combine in the presence of a catalyst to produce hydrogen; however, there is an unfortunate by-product of large amounts of carbon dioxide. Technology has been developed to sequester CO_2 – this entails injecting it into deep wells where it becomes trapped in geological formations. Unfortunately, it is also used in a process called "EOR" or enhanced oil recovery. Here it is injected into old oil wells, which are nearing the end of their useful life. Forcing carbon dioxide down a hole, results in large amounts of oil coming up another hole somewhere else. While this makes sense from a monetary perspective, it isn't really doing the environment any great favors when that oil is burned.

There are some problems at the moment with storing hydrogen that need to be addressed in order to increase performance; however, technologies are steadily being refined, to allow a greater amount of hydrogen to be stored in a smaller space.

However, there are exciting possibilities when it comes to extracting energy from the hydrogen.

Hydrogen internal combustion engine technology

Figure 10.6
BMW Hydrogen7 (hydro-gen ICE vehicle).

When it comes to the point of use, we can burn hydrogen in an ICE. This approach is being used by BMW in their BMW Hydrogen7 vehicle. A large hydrogen cylinder in the boot of the car provides a large engine with hydrogen, which is burned in a spark-ignition engine. This option is attractive to manufacturers, because it allows them to work with a large amount of technology that is existing and they are familiar with. It is also good news to the consumer, because it means that as well as running an ICE engine on hydrogen, a second tank for gasoline can also be engaged, which facilitates filling at conventional gas stations, and extended range. It saves "betting the farm" on one particular type of technology. However, there are some disadvantages. As a heat engine, there is inherently a relatively poor efficiency to the system and, furthermore, because nitrogen from the air reaches high temperatures in the combustion chamber, there are still issues with nitrogen oxides.

Hydrogen fuel cell technology

Figure 10.7
Vauxhall (General Motors) HydroGen3 vehicle.

However, there are alternatives. The fuel cell is one device that could revolutionize transport technology. Fuels cells operate in a manner that is fundamentally different to combustion engines, which burn fuel. A fuel cell doesn't "burn" fuel in a conventional sense; however, it reacts hydrogen with oxygen, using a special membrane, capturing the energy that is released from that reaction in electrical form. The process used to produce electricity from hydrogen is "more efficient" than energy capture from an internal combustion engine, so there is potential to use less energy to do the same amount of work. However, there are some issues with the amount of energy needed to produce and store hydrogen, so this should be borne in mind.

One of the other challenges with moving to a hydrogen economy is that it will require a radical rethink of infrastructure: It will entail the construction of hydrogen filling stations. This is a challenge. No one wants to build a hydrogen filling station unless there are hydrogen cars available to consume the hydrogen, and no one wants to buy a hydrogen car unless there are filling stations available. Undoubtedly, as the questions surrounding this technology are solved, more hydrogen filling stations will be built.

Human-powered vehicles

There is a great case to be made for bicycles (bikes), tricycles(trikes), and other human-powered vehicles (HPVs). A bike can help you get about quickly and cheaply, without the pollution that motorized transport produces. Bikes also take up less footprint on the roads, which means smaller parking spaces for them and less infrastructure. Also, cycling keeps you fit and healthy.

However, there is more to bicycles than just mountain bikes or uprights; there are a variety of vehicles such as recumbent tricycles, which allow you to lie back and relax while cycling, as well as enclosed human-powered vehicles, such as "Velomobiles," which encase the rider in an aerodynamic fiberglass shroud, allowing him to cut through the air more easily. In addition, bicycle rickshaws are becoming a feature of the trendy downtowns of many cities. HPVs also sometimes provide an electrical or petrol assist with a small engine or motor, to produce a vehicle that is human-powered until you reach a large hill, when the assistance will be used to help the rider.

The problem with HPVs is also one of infrastructure. However, they require nothing more sophisticated than a shower and changing rooms at either end of the journey, to allow the rider to freshen up before going on his way.

Figure 10.8
Human-powered vehicle (with electric assist).

 If you are interested in finding more about human-powered vehicles, some good sites to check out are: www.bhpc.org.uk; www.velomobiling.net; www.ihpva.org.

Chapter 11

Engine conversion or fuel conversion?

The choice between going the biodiesel route or the SVO route is basically a choice between converting your engine or converting your fuel; however, the decision is not as simple as that. Running on SVO means you will need to secure a good supply of new oil; running on UCO requires a good supply of used oil and a way of processing it into fuel. Biodiesel does not require any engine conversion but, unless you choose to simply buy it, requires a reaction vessel to make it in, and the storing and handling of some hazardous chemicals.

Of course none of these options are necessary mutually exclusive. You could always have an SVO converted engine that you also run on biodiesel, dinodiesel, SVO, or UCO.

The biodiesel vs SVO argument is a little like the never-ending argument about which are better, dogs or cats. There is no right answer, some people prefer cats, some dogs, it is just personal choice; below are a few questions to ask yourself to help you choose between biodiesel and SVO.

How many vehicles are you running?

If the answer is more than one or two, the expense of converting them all to run on SVO is probably greater than the cost of building, or buying, one reactor to supply all of them with biodiesel. The cost of a kit to convert to SVO is going to be more than US $1000, plus the cost of installation, per vehicle. If you regularly replace your vehicle, you will need to add in the cost of converting each of them to run on SVO to the price you pay for them. Some SVO kits are adaptable to different vehicles, whereas some are very engine-specific, if you are thinking of selling the vehicle as standard, with no conversion, then also consider how reversible the SVO kit you are thinking of installing is and how much it will cost you to change it back.

Case study: Dulas Ltd

Dulas Ltd, a UK firm specializing in sustainable technologies and renewable energy, wanted to ensure that the way that its operatives traveled about to different sites was congruent with its mission of being a firm that treads lightly on the earth, with the minimum possible impact. Because Dulas is fueling a fleet of vehicles biodiesel makes more sense than individual vegetable oil conversions — the business sources biodiesel locally, and stores it on-site in a large storage container (Figure 11.1).

Figure 11.1
Biodiesel storage facility at Dulas Ltd.

The container is fitted with an integral pump, and dispensing equipment, which allows the firm to manage fuel consumption – by keeping regular records and checking they tally with vehicle mileage – and also helps with reporting fuel consumption when it comes to tax and deductions at the end of the year (Figure 11.2).

Figure 11.2
Dispenser allows Dulas to keep track of biodiesel usage for fleet management and taxation purposes.

When an employee wants to use a fleet car, it's a simple job of filling it up on-site using biodiesel (Figure 11.3).

Figure 11.3
Running Dulas' fleet vehicle on biodiesel.

What is your location?

What SVO and UCO do you have available locally and how cold is it where you live? If you live in a cold climate, fuels with a higher gel point are less suitable to biodiesel production because the gel point of your feedstock will determine the gel point of the biodiesel you make from it. However, these fuels are potentially more suitable for an SVO system, because the fuel is heated. However, this all depends on what the gel point is and just how cold it is where you live. Cold filtration can separate SVO/UCO into parts with higher melting points and lower melting points and different feedstocks can be selected at different times of the year.

How are you going to dispose of the waste?

One thing to consider when choosing which technology to go with is what you are going to do with your fuel plant's waste products. Biodiesel production tends to produce a significantly larger quantity of waste than SVO and UCO. With SVO all you have to dispose of are the oily containers that arrived with your oil; but short of throwing them into land-fill, it is a problem hard to find a solution to.

With UCO you will, in addition to oily containers, also have some, maybe quite a bit, of unusable wet or dirty oil to dispose of, along with whatever rubbish you filter out of the oil. Putting this down the drain is not an option. It will put a massive strain on sewerage systems. You are going to need to find a better way of disposing of it, such as composting or getting your local waste oil disposal company to collect it. You could even consider converting it to biodiesel.

Making biodiesel produces a number of waste streams in addition to those mentioned above. You will also have soapy water and methanol-laden glycerol (glycerin); they need to be disposed of. Putting the soapy water down the drain may be an option, but putting the glycerol there too is almost certanly not. See Chapter 12 for further information.

What is your typical journey duration?

If you tend to drive short trips, say of 10 miles or less, your engine is unlikely to get up to temperature to run on SVO/UCO regularly, so you will tend to run it on the dinodiesel/biodiesel tank most of the time, unless you happen to have a vehicle suitable for a single-tank conversion. A lot of two-tank conversions use a small second tank. If you are not using the SVO side of the system, your little tank is going to need refilling very regularly. Still, if you are regularly driving less than 10 miles, then maybe you should reconsider your vehicle use all together: Maybe walking or cycling some of these trips would be a better solution altogether?

Who else uses your car?

Biodiesel is much less of an issue if your car is being used by people other than yourself than an SVO converted one, as they don't have to worry about if the engine is hot enough or switching between tanks. Biodiesel can be used in any ratio with dinodiesel, so fill-ups ought not to be an issue for the other drivers of the car.

Which is greenest?

Arguably, depending on where it was grown, new vegetable oil has a significant environmental and social impact. It is most certainly neither as green nor as ethical as you may be told or want to believe (see Chapter 14).

Biodiesel, when made from *used* oil, has a lower environmental impact than most other options but is still far from totally green. It is made using caustic chemicals and methanol, a fossil fuel. Some people make biodiesel from ethanol and it is possible to make ethanol from nonfossil fuel sources, but this is not realistic for small-scale home production; if you make biodiesel, it is going to be about 20% fossil fuel.

Running on UCO in an SVO converted vehicle is arguably the greenest solution, as there is a lot less in the way of waste to dispose of and no fossil fuel use; however, it *may* be damaging to your engine, which in itself is not very green.

Don't forget that it is almost always going to be greener and cheaper and better for your health to use your bike or walk or catch a train or a bus.

How proven is the technology?

Running on SVO, and especially on UCO, brings with it a greater risk of engine damage than just running on well-made biodiesel. The risks are not generally large and are easily managed with careful consideration, high-quality engine conversion, and high-quality fuels. While a large amount of anecdotal evidence suggests that, when done properly, SVO conversion is safe and reliable, no formal long-term studies have been carried out on SVO or UCO usage in a wide variety of engines.

As for some of the other solutions mentioned in this book, such as mixing your UCO with thinners to lower the viscosity, you must consider these solutions to be at best experiential. Even if in the end no damage occurs, you would have been wiser in the first place to have assumed that you are likely to damage your engine. In addition, you should also consider what exhaust products are produced when your chosen thinning agent is burned.

Chapter 12

Waste streams

We're doing this to save the world right? So what good is it to spend time and effort trying to produce an environmentally benign fuel for your vehicle, if when it comes to tidying up, you tip chemicals down the drain, leave a trail of destruction, empty carboys, and half-junked plumbing fittings in your wake. Biodiesel, if done properly, can *help* with waste management, by reducing the amount of vegetable oil going to landfill – but if done badly, you can pollute watercourses with toxic chemicals, clog up drains, overload sewerage systems, and end up creating ecomayhem. We urge you to make biodiesel responsibly, and keep the planet tidy.

Glycerol

This is the other half of the reaction when we make biodiesel, the by-product. Glycerol is the raw ingredient for soap and all sorts of everyday household items, but unfortunately the stuff we have made is at least 50% not glycerol at all, but a mixture of methanol and lye and soap and water.

Strangely enough, your typical soap maker will not thank you for this smelly mixture. Soap makers are used to getting their glycerol in a nice, pure, clear, odorless, and expensive form.

In order for our smelly glycerol to become the stuff the soap makers want, we need to distill it several times under conditions that are just not achievable at home; we need to find other uses for it.

It is very common to be unsure of what to do with the glycerol by-product and so you end up stockpiling it to deal with when you have worked it out. Don't forget it probably has quite a lot of methanol in it and so any partly full containers of by-product will also be partly full of methanol fumes and that is dangerous. It is probably better that you take it to a landfill rather than accumulating too much of it, but this is a pity as well.

Methanol is, as we know, poisonous to humans, and just about everything else too. Before passing on your biodiesel by-product you must make sure it is methanol-free: Just because we know how to handle the stuff, does not mean we should assume other people do too.

Some books suggest just leaving the by-product outside and the methanol will simply evaporate, but this seems not to be very effective in practice, so you need to boil it off. You can either boil off the methanol in a big cooking pot (well away from you and your house) or you can recover the methanol in some sort of still (see Chapter 7 for more ideas on making a methanol recovery still).

Your biodiesel by-product is a good degreaser on its own; it is good for cleaning up spills and cleaning oily engines and paint brushes. You could turn it into perfectly usable, but funny-looking and funny-smelling soap yourself by first removing the methanol and then looking up soap-making recipes on the Internet. Though what you are going to do with this much soap yourself is another thing, and you may have trouble giving it away to your friends and family.

As we have repeatedly remarked, methanol is poisonous. However, certain critters love eating biodiesel by-product: methanol-munching bacteria and glycerol-munching bacteria. The simplest way to get them to do this is to simply compost the by-product; mixed with card or paper or straw or woodchips, bacteria will break it down and make compost out of it. The problem is that you may end up with a rather large compost heap. Also, some methane digesters can take glycerol (but not methanol) and turn it into burnable gas.

Chemical purification is possible at home using acids; you can obtain up to about 90% pure glycerol at home using this technique. Search the Internet for "acidulating glycerol" for information on using hydrochloric or phosphoric acids.

Glycerol does burn, but if not done at a sufficiently high temperature it will produce acrolein, a very nasty gas.

Soapy water

The water from your biodiesel washes should be reused as far as possible. The water from the last wash is not very soapy and so can be used for the first wash of your next batch of biodiesel. Ultimately, you will still have to get rid of it, however. It is primarily water, then soap, and a bit of methanol and catalyst. Although this solution is not very hazardous, you should not let it escape into the environment down storm drains or onto the ground. Many home brewers put it into the sewerage, claiming that alcohol and lye and soapy water is no more hazardous than vodka and drain cleaner and washing machine water. Methanol is not vodka of course but there are bacteria in the sewer that love to eat it, and disposing of your wash water in this way is not as nasty as a lot of the chemicals your typical Western household routinely puts down the drain.

Methanol recovery is possible with your wash water but not worth the effort; most of your leftover methanol is in the glycerol layer.

If you have space, a gray water system could be considered, such as gravel and reed beds, to clean the water for you before disposal – see The Center for Alternative Technology (UK), www.cat.org.uk, or the solar Living Institute, www.solarliving.org, for books and courses on this.

Oily empties

Carboys, metal cans, or whatever your SVO or WVO came in will start to pile up and can become a major problem to get rid of. Even if the container is recyclable, the recyclers usually don't want to take them because they are all oily. Watch out for anything that used to have methanol or glycerol by-product in it, as these may contain significant amounts of methanol vapor in them. One way of disposing of oily containers is to simply landfill them, which is a pity when they are recyclable. You will have to clean them before recycling them that is not easy either.

Remember your three R's: reduce, reuse, and recycle. Reduce your need to dispose of the containers by not collecting them at all. If you can't do that, then reuse the containers next time or find another use for them. If you can't do, that, then recycle them … and landfill them only as a last resort.

Chapter 13

Health and safety

Boring–boring–boring, yawn ... "*Health and safety – Health and shmafety.*" We can hear you now ... skip this section and get on with the good stuff? However, this is possibly the most important section of the book!

Life is full of risks and a risk-free life would be a pretty dull one (ask an insurance salesman what he does during the weekend?). However, most of us don't like to expose ourselves, our friends or pets or homes or the wider environment to unnecessary or extreme danger.

Biofuel production, and in particular biodiesel, is no exception. Although some of the processes and materials are potentially very dangerous, if we are sensible, careful, and thoughtful about what we are doing, the risks can be managed.

It is not about removing the risks – that is both impossible and impractical – but we can seek to minimize them and implement simple control measures where possible to ensure that the risk to ourselves and others is minimized.

Heavy things and manual handling

The construction of biodiesel processing equipment can involve the manipulation of some large and bulky things: it's easy to underestimate the weight of tanks, with only a little oil in them, not to mention when they are full. Even a 5-gallon container full of oil could put your back out if not lifted carefully.

Humans may consider themselves the most highly evolved creature going, but it was not so long ago, in evolution terms, that we started walking about on two legs. It was a great moment for us, as it freed up our hands to make biofuels, but our backs have not quite caught up, they are still better suited to walking about on all fours and are easily damaged.

Ensure that you maintain a good posture when moving large items, and also that you are capable of moving the item in question. Bend your legs, not your back, and don't twist or turn sharply.

When lifting large or heavy objects, keep your feet facing straight and shoulders wide apart, to give you a broad, stable base – this will ensure that you are less likely to slip.

Seek assistance from a friend, or consider using lifting equipment, such as a hoist or engine crane, if you are not capable of lifting tanks yourself. Remember, there's always another day to move a tank, but you're only going to get one spine, so don't damage it in the pursuit of green fuel.

Working with/under vehicles

If you decide to perform a vegetable oil conversion on your vehicle, you are going to require access to part of your car that you just can't reach easily. This may involve getting under your vehicle or jacking your vehicle up. If you are going to jack your vehicle, ensure that you invest in a proper set of axle stands, rated to support the weight of your vehicle. Piles of paving slabs or bricks are not the solution; if they crack, the car will fall on you, and it will hurt like hell if you're lucky and kill you if you're not. Ensure if you are just jacking the front or rear of your vehicle that the other wheels are braked, and adequately supported with chocks to prevent the vehicle from rolling. If you're going to work in a confined space under your vehicle, ensure that there is someone there to watch and ensure your safety. It's always good to have a buddy there when something goes wrong.

Ensure that you can slide in and out from under the vehicle easily. If your joints creak a bit and shimmying yourself in and out is a bit of a chore, consider investing in a roller skid to allow you to slide under your vehicle easily. It will raise your back off the floor, preventing the cold garage floor from making it stiff, and will ensure that you can get in and out from under the vehicle quickly.

Good housekeeping

Many accidents can be avoided by tidying up, ensuring a clean working environment, and just general good housekeeping – a pile of oily rags can soon turn into a fire; loose tools and sockets on the floor can soon turn into a slipping hazard, causing you to trip or fall. Ensuring that you work in a tidy space will prevent a lot of accidents.

Biodiesel risks

Biodiesel production involves some nasty chemicals, but none of them need to be feared. Simply treat them with respect: Methanol is flammable and poisonous; lye/catalyst is caustic.

Methanol has two forms that we need to worry about: It can be liquid or a vapor; both states are poisonous and flammable. It is both colorless and odorless and, as a vapor, it is heavier than air and will collect on the floor or in hollows. Most methanol accidents occur as a result of carelessness or storage; be careful and thoughtful when handling it; try not to buy more than you need; and try not to store it.

Poison

You don't want to be getting methanol on your skin, in your eyes, or in your mouth, and you don't want to be breathing it in:

- Whenever handling it, wear sufficient clothing, long sleeves, long trousers, proper shoes, an apron, gloves, and goggles.
- Remember, even expensive vapor masks offer no protection against methanol fumes.
- Always have adequate ventilation, active ventilation (fans and extractors) are preferable.
- Vent your reactor to outside your workspace.
- Don't lean over any open containers.
- Don't leave any containers open.

Flammable

Methanol burns with an almost invisible flame and, if in a confined space like a tank, is explosive:

- Never smoke anywhere near methanol.
- Try not to buy more methanol than you need; try not to store it.
- If you do have to store methanol, be very careful where you store it – not near your house and not in the sun.
- Vapor will accumulate in "empty" containers, like your methoxide mixer or your reactor or containers half full of your biodiesel's by-product/glycerin.
- Sparks from electrical switches and motors will ignite fumes; don't allow fumes to collect and only use explosion-proof equipment.

If you get methanol or methoxide on your clothes, you will need to get the clothes off as quickly as you can. However, pulling a methanol-soaked shirt over your head and face is not a good idea: Cut it off instead. Wash any methanol spills with plenty of water.

Methoxide, methanol, and lye should be treated as a caustic burn and washed down with copious amounts of water for at least 1 hour; see the caustic burns section below.

Fire

Obviously methanol is flammable! The fact that it becomes a vapor at a relatively low temperature means that you need to be especially careful with naked flames or other sources of ignition around methanol.

Oil, especially hot oil, is flammable. We've all seen the videos of chip-pan fires, and the damage that they can cause – remember in our biodiesel processing operations, we are using a considerably greater quantity of oil than is found in a conventional chip pan. Make sure you are familiar with the correct methods of putting out different types of fire. Make sure that there is a selection of fire extinguishers around, and that you are familiar with the different types of extinguisher: Know which one to select, depending on the type of fire, and know how to operate them in an emergency. Ensure that you also have some buckets of sand and a fire blanket around – all are necessary if the proverbial hits the fan.

Rags soaked in oils can spontaneously combust; it is rare, but it does happen. Don't leave them lying around. Ensure that you dispose of any soiled or oily rags correctly and store them away from sources of heat and ignition.

Fire extinguishers are a good thing to have about: Foam or water ones are not very useful to you; use CO_2 or powder ones. Keep one by the door, so it is accessible when you're arriving or leaving the building, and don't use it as a door stop, which is illegal in a lot of places!

A Class F-type fire extinguisher for dealing with wet chemical fires is the best one to have around for homebrew biodiesel. This fire extinguisher uses a heavy alkaline solution to "saponify" the oil (turn it into soap!) Make sure you use an extinguisher designed for Class F fires. Other fire extinguishers not designed for oil fires will tend to just spread the oil around and make the problem worse.

Practical safety hints for biodiesel production

On a practical note, there are a lot of things you can do when making your biodiesel to keep risks to a minimum and ensure that if things do go wrong, you've got the correct gear around to help you sort out the problem.

Some tips:

- Always have eyewash and running water available.
- Don't use plastic reactors.
- Deal with leaks promptly. They are a fire hazard, and a slip hazard, and may be a sign the equipment is failing.
- Don't use blenders to make test batches; despite what it says in some books, they are not designed to cope with the chemicals and don't have spark-proof motors.
- Watch out for plastics in your reactor degrading, especially any hoses. They tend not to be chemical resistant and may spring a leak, possibly spraying you with hot oil and methoxide. Try not to use plastic hoses; replace with metal if you can, and if you do use them, then replace them before they become discolored and hard.

- Biodiesel or waste oil itself is not very hazardous, not very flammable, and not very poisonous. Nevertheless, it is not a good idea to drink it and you should be careful where you store it and what you store it in and make sure it does not escape into the environment.

The Collaborative Biodiesel Tutorial has excellent healthy and safety information about making biodiesel, especially with regard to the Appleseed and similar reactors: www.biodieselcommunity.org/safety/

Electrical safety

You will be connecting pumps and heaters to a mains electrical supply. When mains electricity mixes with fluids and people, the results are seldom a laughing matter, so it is important to make sure that you are competent to perform the wiring that you are going to take, and if you do not feel confident you seek appropriate assistance.

It is important that the fixtures and fittings that you are using to wire the motors and heaters of your biodiesel processor are correctly rated for the task in hand. Immersion heaters draw a large current, so it is important that you use the correctly sized wires to connect them up, else the cables will overheat and you are at risk of creating a fire hazard.

Also, you need to ensure that all of the fixtures that you are using are "ingress protected," also known as "IP sealed." Ingress protected fixtures are ones that are designed with special grommets, gaskets, and seals to prevent the ingress of solid matter (dust and chemicals) or moisture from fluids. Where a cable enters or exits the enclosure, suitable hardware is provided to ensure that a watertight seal is created.

When working with mains voltage, it is sensible to use a circuit breaker rather than an "old school" fuse. Circuit breakers react faster to a faulty condition. When a fault is created, the chances are the current will return to earth through you, or if you're unlucky the kids or pets. In this situation, fractions of a second are vital. A miniature circuit breaker (MCB) will react quickly, protecting people, pets, and property, whereas a fuse will take a longer time to blow or may not blow at all.

There are different types of circuit breakers, that serve different functions. An MCB operates on the principle of overcurrent – once a certain current has been exceeded, the breaker will trip – whereas an RCD – a residual current device – operates on a slightly different principle, monitoring the live and neutral (or hot and ground) to see if any current is going "missing" while traveling round the circuit. If the RCD senses more current going "in than out," then it will trip – this is what happens when an earth fault arises.

It is best to have both kinds of protection on the circuit you use to power you biodiesel reactors. Overcurrent will protect against short circuits, whereas RCDs will protect against any live component making a circuit to earth, which may well be via you.

Slippery oils

Oil is slippery stuff. Oil on the floor can cause you to slip over, whereas oil on your hand or on a container can cause you to lose grip of it. Making biodiesel and processing waste oils is going to be a messy occupation and it is essential that you clear up after yourself by soaking up any spills with rags or sawdust or even special oil-absorbing granules.

Heat

Whether just dewatering waste oil or making biodiesel, heat is usually required. Making biofuels we often end up heating large quantities of oil to around 50°C (122°F), maybe more, and care needs to be taken to avoid burns and scalds and fires.

- Never ever (ever!) use a flame to heat your oil; electrical heaters controlled by a thermostat or an indirect heating method must always be used.
- Never leave your oil while the heater is on, even just for a moment.
- Have a fire extinguisher handy, and make sure it is the right type; water is no good for oil fires.
- Have running water available in case of burns and scalds.

 ## Chemical burns

Lye, NaOH, or KOH is nasty stuff; it is caustic, which means that it burns. It will take moisture from around it, from your skin or from the air, and will do you some nasty damage. What is more, you may not feel it as it kills the nerves as it goes; some of the nastiest caustic burns are on the feet where the person thought the problem had gone away and could not see the damage as it continued inside their shoes.

Vinegar is not effective against caustic burns. The idea that it is stems from the theory that one needs to neutralize the alkali, the lye, with an acid, the vinegar; but vinegar is a very weak acid and lye a very strong alkali. The correct treatment for alkali/lye burns is to flush the area with copious amounts of water, see below.

Preventing caustic burns

Prevention is always better than cure, right! Keep caustic chemicals out of harms way; keep them in a locked cabinet if necessary. Wear protective clothing whenever handling lye or methoxide: always use your goggles and your chemical-proof gloves, proper shoes (no sandals, hippie), a dust mask, long sleeves, long trousers, maybe an apron.

Treating caustic burns

The first thing when tackling any emergency is not to become a victim yourself. Before helping someone else, put on adequate protective clothing yourself.

If you have spilled lye on your shirt, don't pull it over your head; cut the shirt off, otherwise you could transfer caustic chemicals to your face or eyes, making things much worse.

- **Flush the affected area with copious running water for *at least an hour* and seek medical attention.**
- **If the eyes have been affected, continue irrigating the eyes until medical help arrives.**
- **Do not use vinegar (see above).**
- **Do not apply any ointments to the burn area.**

Graham Laming's page on caustic burns is excellent and thoroughly explains why vinegar is not a good treatment for caustic burns: www.graham-laming.com/bd/first_aid_caustic_burn.htm

Chapter 14

Biofuel ethics

Introduction

Biodiesel is a subject that currently polarizes the renewable energy community; some hate it while others love it. Initially viewed as a "magic bullet" solution to overcome the problems with the carbon emissions from fossil fuels, the tide has quickly turned against biofuels as a sustainable solution.

There are good arguments on both sides of the fence, but at a time where we desperately need solutions to the impending crisis we face, brought about by resource scarcity and climate change, there are tough questions to be asked about whether it is appropriate to look to technologies to provide mobility, which compromise our collective ability to feed ourselves.

The carbon cycle – the bike with the wonky chain

We explored briefly in Chapter 1 the fact that the "carbon cycle" with biofuel production isn't perfectly circular as some would like you to believe; but instead has inputs of additional carbon at every stage of the supply chain. It is often hard to quantify the "carbon content" of biofuel; you have to know the carbon intensity of the processes used to grow, harvest, and process the crops, and to produce pesticides and agricultural chemicals used at each stage of the process.

The main thing to remember, is that the more "inputs" to the process that you can remove, the less carbon you are inputting into the process. However, this can have consequences too. If for example, you manufacture your crop organically you will be omitting or reducing inputs of agrichemicals, but the result of this is that farming to produce the same amount of yield may require an increase in work and labor. If you are making your biofuel from something that would otherwise be disposed of as waste, such as used cooking oil, then you are reducing the burden on "waste sinks," the embodied carbon in the used oil would otherwise be wasted.

Interdependence, not independence

Many sources have cited how biofuels are great for national "fuel security" by reducing dependence on other nations for oil. While they can help the problem, at our present levels of fuel consumption we cannot hope to produce enough biofuels using current generation technologies, even if we diverted all our agricultural resources into producing biofuels – as we would then have to find some other way of feeding ourselves, presumably trading "fuel security" for "food insecurity."

In a globalized world where your clothes are made in Italy, your TV made in Japan, your food grown in South America, and your trainers made in Africa, we need to accept the reality that with globalization, we are more dependent on each other than ever before, and that notions of "energy independence" are a little misleading.

Can we grow enough biofuels?

In short, no. There is not enough global land area for us to grow enough biofuels to meet our present volume of fuel consumption. Ultimately, this means that in the long term we need to look for solutions above and beyond what liquid hydrocarbon fuels provide for us. There is some possibility that biofuels can help ease the transition as oil supplies dwindle and we develop alternatives; however, in the long run biofuels are not going to power us into the future. They are, at best, part of a stop-gap solution while we ease our dependence on fossil fuels.

▲ Figure 14.1
Oilseed rape crop. Courtesy: CNH UK Ltd.

However, current methods of producing biofuels, known as "first-generation biofuels," could be superseded by new improved methods of production, second- and third-generation, which have the potential to produce far greater yields of fuel.

Biofuels, global warming, and big business

For a long time we thought that biofuel production was going to produce less net carbon emissions than burning the equivalent fossil fuel, but it turns out that this was both optimistic and naive.

One of the big problems with the hasty growth of the biofuels sector, driven by the high value of biofuels crops, is that unscrupulous companies are clear-felling large areas of tropical rainforest in order to grow oil crops like the oil palm, partly to supply us with biofuels. In September 2005, Friends of the Earth published a report into the impacts of palm oil production. "Between 1985 and 2000 ... the development of oil-palm plantations was responsible for an estimated 87 per cent of deforestation in Malaysia."[1] In Sumatra, Borneo, Malaysia, and Indonesia tens-of-millions of hectares of forest have been converted to palm farms and thousands of indigenous people have been evicted from their lands, as George Monbiot points out "The entire region is being turned into a gigantic vegetable oil field."[2]

It is not just lost rainforest, its species and its people who are suffering; deforestation accounts for 25% of all greenhouse gas emissions,[3] the loss of biodiversity and the loss of some of the world's most valuable carbon-sinks is very bad news for the environment; it is no exaggeration to say that this method of biofuel production is actually worsening the effects of global warming and climate change, all in order to produce biofuels that well-meaning consumers buy at the pump thinking they are doing the good thing by choosing a "sustainable alternative."

Food vs fuel

If we accept that there is a finite amount of land dedicated to agriculture, and we do not encroach on virgin habitats or environmentally sensitive areas, then we are faced with a stark choice of food vs fuel. Even if we start devoting a proportion of agricultural production to energy crop and biofuel production, we are competing with food production and causing the price of food to rise.

While rich Western consumers are cushioned from this price increase by their affluent lifestyles and high wages (the average Westerner spends around 16% of their income on food), the hardest hit are the poorest, where prices of basic food commodities increasing by only a small margin can mean life or death to those living on less than a dollar a day.

Professor John Beddington, the UK government's chief scientist, has called food shortage as a result of biodiesel production the "elephant in the room"[4] and seriously questioned both the ability to produce any meaningful amount of fuel using this technology and the British Government's current stance on biofuels.

It's not just the land that biofuels compete for, it is also agricultural subsidies, fertilizers, and agricultural chemicals (which are usually made from fossil fuels anyway); investment, infrastructure, water, and labor are also put under increased strain.

However, some argue that this is an oversimplification and that we need to view the world food situation in the context of the ever-growing demand for food to provide feed for animals and to meet the growing demand for meat in countries like India and China, which in turn is also driving the surge in prices of arable crops. Furthermore, the rising price of crude oil is impacting both the cost of production of food, and price of transporting it.

In the context of a world where 854 million people have insufficient food to eat[5] the enormity of the problem cannot be underestimated, rising population will necessitate a 50% increase in food production by 2030,[6] we must not allow our desire for fuel to come before others' need for food.

Biofuel from waste oils

As we have seen, biodiesel can be produced from waste vegetable oils from the food industry. Fast food and other catering outlets provide used "grease," which up until recently has been used in the manufacture of animal feed. However, since BSE (mad cow disease), used cooking oils have been banned from use in the production of feed, this changed the "grease" from a valuable resource into a waste product, which had to be disposed of by land-fill. Unfortunately, there are limits to the quantities of used oil available, the British Association for Biofuels and Oils state that biofuels produced from used oils could only meet 1/380th of the diesel fuel requirements in the UK.

Biofuel from agricultural waste

Second-generation biofuels, production of fuel from lignocelluloses (the woody part of the cell walls of plants), has the potential to produce more than twice as much fuel as is currently being produced from the waste products of the agri-food industry. Where grain might be used for making flour, the stems and the chaff could be processed to produce liquid hydrocarbon fuels. Furthermore, the overall impact of producing the fuel is vastly reduced, as it is making use of the waste products from the production process, rather than the "prime" food component of the crop.

Fuel from algae

Often called "third-generation biofuels," algaeculture can produce vastly more biofuel per hectare than conventional crops. The technology is still in its infancy but, despite a few worries, appears to have great potential, with dozens of new companies currently promising great things; see the Epilogue for more on this.

Other alternative transport technologies

When considering the optimum technology to deliver our (auto)mobility needs, we need to consider the amount of energy that we can produce from any given land area. Biofuel production from conventional methods results in a meagre production of energy and a small number of "miles driven" produced from any given land area. Technologies such as wind power coupled with electric vehicles and a variety of solar technologies have the potential to deliver a much greater return of miles for any given land area.

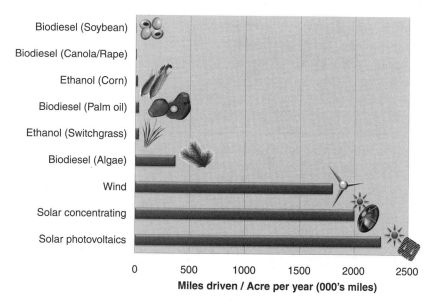

▲ Figure 14.2
Comparison of some alternative transport technologies by acres/mile.

One criticism often levied at alternative vehicle technologies is that they cannot produce the same "range" of travel as internal combustion engine vehicles that use liquid hydrocarbon fuels. At present, to some extent, this is true; the range of even the best electric vehicle is significantly less than that of even a modest gasoline or diesel vehicle. However, when we compare our needs with our expectations, we find that for the average driver alternative vehicles can meet the needs of most of our everyday journeys.

Part of the problem is societal, we have come to expect that a single vehicle will meet all of our transport needs; however, if we consider using one vehicle for our day-to-day driving, and then, for example, renting a truck when furniture needs moving, walking or cycling short distances, travelling long-distance by public modes of transport, or renting a traditional long-range vehicle when special needs occur; then we can overcome the majority of these problems. The solution here is not a more efficient fuel technology but a more efficient way of thinking!

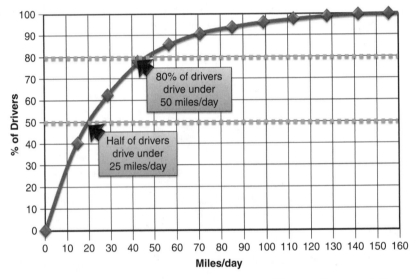

Data source: US dept. of transportation,
Fed. highway admin.
Nationwide personal transportation survey

▲ Figure 14.3
Miles/day vs percentile of drivers.

Conclusions

As a transitional technology away from fossil fuels, biofuels offer limited potential to produce liquid hydrocarbon fuels for applications where a transition to other technologies is not yet possible. However, large-scale production of biofuels is already having, and will continue to cause, enormous impacts affecting issues of food supply, equity, deforestation, and climate change.

There are strong arguments in favor of making biofuels from waste oil but there is also an irony in that in doing so we stimulate interest in biofuels and so create a demand for them. This in turn creates a new market for big business to exploit, and business is of course motivated by profit. Big business will find, indeed already is finding, cheaper, less ethical, supplies of the materials it needs.

While the arguments currently raging in the media are somewhat oversimplified, and to some degree biofuels are being made a scapegoat for the increase in world food prices (which are influenced by a much more complex combination of factors), first-generation biofuels are no panacea for global warming, climate change, or the rising cost of fuel. Second- and third-generation biofuels may hold some hope, and in the Epilogue you can read about some future directions for biofuel.

References

1. "Friends of the Earth." The oil for ape scandal: How palm oil is threatening orangutan survival. Research report, September 2005. www.foe.co.uk/resource/reports/oil_for_ape_full.pdf

2. Monbiot, G. "Worse than fossil fuel." *Guardian*, December 6, 2005. Accessed online: www.monbiot.com/archives/2005/12/06/worse-than-fossil-fuel/

3. Rainforest Action Network. Call for an Immediate moratorium on US incentives for agrofuels, US agroenergy monocultures and global trade in agrofuels. Accessed online: ga3.org/campaign/agrofuelsmoratorium

4. Smith, L., Elliott, F. "Biofuels threaten 'billions of lives.'" *The Australian*, 2008.

5. WFP (World Food Programme). "FAO & the state of food insecurity." Accessed online: www.wfp.org/aboutwfp/facts/hunger_facts.asp.

6. Smith, K., Edwards, R. "The year of global food crisis." *Sunday Herald*, 2008.

Further reading

Adam, D. "To bio or not to bio – are 'green' fuels really good for the earth?," *The Guardian*, Saturday, January 26, 2008.

AFP. WFP chief warns EU about biofuels, 2008.

"British Association for Biofuels and Oils." Memorandum to the Royal Commission on Environmental Pollution. Accessed online: www.biodiesel.co.uk/press_release/royal_commission_on_environment.htm.

"Greenpeace." A bad day for the climate as biofuel legislation kicks in. Accessed online: www.greenpeace.org.uk/tags/biofuels.

Neal-Stewart, C. Jr. To the editor: "Biofuels & biocontainment." *Nature Biotechnology*, 2007; 25(3).

Marvey, B. B. "Fats & oils: why all the fuss?." *Science in Africa*, 2002.

Monbiot, G. "Feeding cars not people," *Guardian*, November 22, 2004. Accessed online: www.monbiot.com/archives/2004/11/23/feeding-cars-not-people/.

"Union of Concerned Scientists." *Biodiesel basics.* Accessed online: www.ucsusa.org/clean_vehicles/big_rig_cleanup/biodiesel.html.

WFP (World Food Programme). "Food as aid: trends, needs and challenges in the 21st century." Accessed online: www.wfp.org/aboutwfp/facts/hunger_facts.asp.

Chapter 15

Biodiesel and vegetable oil are not just for cars!

Introduction

Diesel engines are ubiquitous in many applications, cars, trucks, buses, and commercial vehicles, all of which make use of the diesel engine's high torque and superb reliability. We're going to explore some different applications where biodiesel or SVO can be successfully employed.

There are lots of things we "can" do, but not all of them are desirable. In a post-oil society, liquid hydrocarbon fuels will be very scarce. In Chapter 14 we explored why we can't produce enough biofuels to meet all our transport needs.

Some vehicle applications will benefit from Battery electric vehicle and fuel cell technologies, reducing the need for liquid hydrocarbons; however, in applications such as aviation, for example, the unique energy-dense properties of the fuels we have grown used to makes them hard to replace. It is hard to envisage an airliner powered by batteries because of fundamental limitations of the technologies – they are just too heavy.

What comes out of this is the realization that we have a "hierarchy of energy needs;" some applications can be met by a wide range of fuels, whereas others require an energy source that fits a specific set of criteria.

At the other end of the scale, we can make heat relatively easily from burning wood and solid fuel, so generating "heat" from burning biodiesel is a waste of a scarce fuel for a specialized application.

We all have different worldviews regarding what is important to us to maintain an appropriate standard of living. We need to reconcile our needs against our wants, and find suitable technologies for suitable applications, if we are to develop lasting solutions that will take us beyond the end of oil.

Motorbikes

EcoRider supplies off-road diesel motorbikes, which obviously can be run on biodiesel too. Made in Scotland, they have a tiny 230 cc diesel engine, 10-inch wide tires – and can travel at up to 25 mph. Such a small compact solution is very frugal with the amount of fuel it uses – reducing the amount of fuel needed in the first place.

www.ecorider.com

Royal Enfield of India produces the classic British motorbike but sadly it no longer produces the diesel version. Hayes Diversified Technologies produce the M1030M1, the world's first diesel-powered motorcycle designed specifically for military use. And there are others too, some made in the Ukraine, some still in development. The following websites have more information on diesel-powered motorbikes:

www.journeytoforever.org/biodiesel_bikes.html

www.dieselbike.net

Boats

New Zealander Pete Bethune circumnavigated the globe in his 78-foot Earthrace trimaran. The boat, designed by Craig Loome Design, uses two 540 hp Cummins diesel engines fueled by biofuel produced from a range of feedstocks. Pete decided to show that biodiesel could be produced from a versatile range of feedstocks by having a pound of fat removed from his own backside and used in his vehicle fuel. The boat aimed to usurp the British Boat "Cable and Wireless" from its record-breaking time; however, the team ran into a number of difficulties and had to abandon the record attempt.

Airplanes

The emissions from the aviation sector globally are one of the faster growing areas of CO_2 emissions and, unless you never fly anywhere, almost certainly form the greatest proportion of your climate-changing-gas emissions. To put this another way, the chances are that cutting out flying altogether is, for most people, both an incredibly effective and a very simple method of massively reducing your carbon footprint.

Green Flight International started by conducting tests on blends of aviation fuel and biodiesel, eventually working up to blends of 100% biodiesel fuel in their jet aircraft. They proved successfully that biodiesel is good to go at 17,000 feet. In terms of performance, the biodiesel was found to perform as well as traditional aviation fuel.

Green Flight International

www.greenflightinternational.com/

A Virgin Atlantic jumbo jet also proved that biofuel could work in commercial-size jets, when Virgin flew a 747 jumbo jet, with one of the jet engines connected to an exclusive biofuel feed, contributing 20% of the engine's power. The fuel used was made from a combination of coconut oils and oil from the babusso nut from Brazil. There are concerns that biofuels are more likely to freeze at high altitudes.

Portable generators

If you are trying to live an off-grid lifestyle, the chances are you have considered an array of renewables options for your land. For places where the grid is inaccessible, an off-grid setup of renewable devices – microwind, microhydro, and solar power – can meet many modest electricity needs. Battery back-up is a good way to ensure that energy you generate in times of plenty can be saved for times when the sun is hidden behind the clouds and the wind is still; however, there will always be times when you need to run a bigger load, or need some supplementary power. A biodiesel or SVO conversion fired generator provides a perfect solution for back-up power.

Chapter 16

Epilogue

In this book, we've attempted to cover the current state-of-the art for the "home" biodiesel producer or SVO converter. If we look to the past couple of years of hobbyist biodiesel production as any sort of indication for the future, one thing is clear: As time goes on, hobbyists have become more professional, and improved their processes. In the very early days it was all open tanks and simple devices; now more sophisticated home reactor designs are evolving, more sophisticated control processes, and the hobby itself is moving forward as we collectively learn and share information.

There are some interesting directions for commercial biodiesel that could yield more sustainable ways of making fuel from biological matter; however, it is unclear at the moment as to whether any of these processes will be in the reach of the home hobbyist. Things that sound crazy today may very well start to look viable as the price of oil begins to soar.

The kids from NERD?

Next-generation biofuels are set to deliver fuels that can produce greater yields of biodiesel per hectare, and potentially offer a more sustainable alternative to the present methods used for making biofuel (FAME). Where FAME stands for fatty acid methyl esters (for an explanation of this acronym, see Chapter 5), NERD stands for non ester renewable diesel. If you've read the stuff in Chapter 5, you will know that esters are the bedrock of homebrew biodiesel, so second-generation biodiesels will employ fundamentally different technology.

The industry is currently struggling to get to grips with next-generation biofuel technologies. There are still challenges ahead; however, the technology shows a lot of promise. The first NERD refinery was opened by Neste Oil in Porvoo, Finland, in summer 2007.

Algae

For some time, biofuel produced from algae has been seen as something that is on the horizon; however, a number of companies have made significant inroads to commercializing this technology. Petrosun switched their plant over to commercial operation on April 1, 2007, with over 1100 acres of open ponds. In Texas it is a cheap method of producing algae; however, open ponds afford less control over how the aquatic ecosystem develops. Meanwhile Green Fuel Technologies have announced that they are ready to build a commercial-scale plant.

PetroSun: www.petrosuninc.com

Green Fuel Technologies: www.greenfuelonline.com

Algael biodiesel offers a number of exciting opportunities. First, algae is very cheap to grow: It just multiplies in water with little encouragement, and is a very efficient means of coverting solar power into "energy." Unlike expensive "engineered" solutions to automobility, like solar photovoltaics powering advanced electric cars, algael biodiesel has the possibility of being a very cheap means of providing transport fuel, using relatively basic technology.

There is also the possibiliy of using algae to capture carbon dioxide from existing power plants, as algae takes in carbon dioxide for its growth. Sure this CO_2 gets released straight back into the atmosphere when the fuel is burned, but the ability to capture it makes the process of growing algae more efficient and sequesters the carbon (albeit *very* temporarily) before it is released back into the atmosphere.

However, aquaculture does not differ from agriculture when we look at the impacts of large-scale monoculture cultivation.

Nanotechnology in the production of biodiesel

Nanotechnology is a word that is firmly rooted in the futurologist's lexicon. By engineering on a small scale, we can make much more efficient use of materials and do things with less energy (which is good for efficiency, and good for the environmental impact of our processes).

Victor Lin from Iowa State University, is hoping to revolutionize biodiesel production, by using nanotechnology to make the process more efficient and less environmentally damaging. Victor has chosen to focus on the catalysts used in the process of making biodiesel, rather than using the simple lye (base/alkaline) catalyst that we focus on in this book (which is the current industry-standard method); Victor plans to use "nanospheres."

If you look back to Chapters 5 and 6, you'll see that free fatty acids (FFAs) are a big problem in biodiesel production. Victor's nanospheres are loaded with a basic catalyst, which performs the same function as the lye in our simple reaction, and acidic catalysts to react with the FFAs.

This takes methoxide out of the equation entirely, and removes a number of processes from biodiesel production, potentially making the process more efficient. It also makes handling the catalyst much easier, as the nanospheres are solid, meaning that the catalyst can be easily recovered from the solution.

The process also results in cleaner, more uniform output products, with better-quality glycerol as a by-product, and better-quality biodiesel.

The technology currently works in the lab, and with investment from venture capital, Victor hopes to scale up the process to 300 gallons per day.

Final remarks

Our hope for the future of biodiesel and renewable fuels is that the debate and argument around them grows in sophistication. It's not black and white, and in a world that is facing dwindling reserves of oil, it is far too early to throw any sustainable technologies out of the pram and discount them.

Biodiesel is one of a range of solutions that need to be considered: Each has its pros and cons, and each technology will reach maturity at a different time. To stabilize the level of atmospheric carbon dioxide *and* work within the constraints of resource limitations, it is clear that there is going to be a need for a lot of innovative thinking and room for a range of solutions in the years ahead.

Technology will undoubtedly advance, and the hobbyist community, as ever, shows its amazing adaptability to change, innovative thinking, and collective creating problem community. What differentiates hobbyists from big business is that, by-and-large, hobbyists are motivated by trying to do the right thing, and achieving concrete change, through small steps. It's the people (the names in the Acknowledgments are a good place to start) who advance the state of the hobby step by step that make the biodiesel community really special.

It's an exciting journey, but recapping our thoughts in Chapter 1, it's not where you're going, or what route you take, but the sustainability of the fuel you put in your tank to get you there.

Appendix

Abbreviations and acronyms

B5   a mixture of 5% biodiesel and 95% petrodiesel

B20   a mixture of 20% biodiesel 80% petrodiesel

B100   100% biodiesel (0% petrodiesel)

CNG   compressed natural gas

CO   carbon monoxide

DERV   diesel engine road vehicle; normal UK, duty/tax paid, road diesel fuel, as opposed to "red diesel"

DfT   Department of Transport

DOE   Department of Energy

EPA   Environmental Protection Agency

FAME   fatty acid methyl ester

HC   hydrocarbons

MeOH   methoxide (an abbreviation – NOT a proper chemical formula)

MVO   modified vegetable oil

NO_x various oxides of nitrogen

OSR   oil seed rape

PPC   pollution prevention control

RME   rapeseed methyl ester

SVO   straight vegetable oil

ULSD   ultra low sulfur diesel

UVO   used vegetable oil

WVO   waste vegetable oil

Biodiesel courses and classes and workshops

In the UK, there are many courses to teach you how to make biodiesel; unfortunately, most of them are thinly veiled promotional exercises designed to teach you how to use their product. Below are a few we recommend:

Courses in North America

Maria "Girl Mark" Alovert's excellent courses
www.girlmark.com

Solar Living Institute, Hopland, CA
www.solarliving.org/workshops/

Courses in Europe

The Centre for Alternative Technology runs biodiesel and SVO courses:
Convert your engine to vegetable oil (3 days)
Make your own biodiesel (3 days)
www.cat.org.uk/courses/

Low-Impact Living Initiative
www.lowimpact.org

Sundance Renewables

A worker-owned not-for-profit cooperative in South Wales that does an excellent course for groups entitled "Biodiesel Production as a Community Enterprise" for those wanting to set up their own plant:

www.sundancerenewables.org.uk/biodp/training%20course.htm

Warranty statements from engine/vehicle manufacturers

If you are interested in converting or running your vehicle on biodiesel, it is wise to have a look at your vehicle manufacturer's warranty statement first of all. Here, we provide web links to vehicle manufacturers' warranty statements concerning the use of biodiesel in their vehicles and engines.

Be aware that the manufacturers listed will be warranting their engines for use with commercial-quality biodiesel, produced and tested to exacting standards. The biodiesel that you produce at home is unlikely to conform to these standards, so any concession on the warranty should serve as an "indicator" as to the suitability of your engine for use with biodiesel, not a guarantee that the manufacturer will pick up the pieces if all goes wrong.

Under US Federal law, the Magnuson Moss Warranty Act, your engine's warranty cannot be invalidated simply because of the use of biodiesel, even if the manufacturer does not recommend its use. That is to say that they cannot void the warranty simply because you used biodiesel; they have to prove that any problem is down to the fuel, whether that be biodiesel or anything else, at which point it becomes the responsibility of the fuel supplier. If any problem is down to faulty parts or bad workmanship then they must honor the warranty regardless of what fuel you have been using.

Case IH

- www.caseih.com/highlights/highlights.aspx?&navid=121&RL=ENNA&typeid=6471803&recordid=151
- www.biodiesel.org/pdf_files/OEM%20Statements/20060724_Case_IH.pdf

Caterpillar (CAT)

www.biodiesel.org/pdf_files/OEM%20Statements/2005_OEM_CatVersion9.pdf.

Cummins

www.biodiesel.org/pdf_files/OEM%20Statements/2004_OEM_cummins.pdf.

Detroit Diesel

www.biodiesel.org/pdf_files/OEM%20Statements/2005_DDC_Statement.pdf.

Ford

www.biodiesel.org/pdf_files/OEM%20Statements/2004_OEM_ford.pdf.

General Motors

www.biodiesel.org/pdf_files/OEM%20Statements/2004_OEM_gm.pdf.

International

www.biodiesel.org/pdf_files/OEM%20Statements/2005_May_OEM_International.pdf.

John Deere

www.biodiesel.org/pdf_files/OEM%20Statements/2004_OEM_john_deere.pdf.

Mercedes Benz

www.biodiesel.org/pdf_files/OEM%20Statements/20060608_Mercedes_Benz_bio_position.pdf.

New Holland

www.newhollandmediakit.com/index.cfm?fuseaction=newsreleases.DisplayNewsReleases&NewsID=166.

UD Trucks/Nissan Diesel

www.biodiesel.org/pdf_files/OEM%20Statements/2006_Nissan.pdf.

Volkswagen

www.biodiesel.org/pdf_files/OEM%20Statements/2005_OEM_VW%20US%20Biodiesel_Statement_5_16_05.pdf.

Volvo

www.biodiesel.org/pdf_files/OEM%20Statements/2005_Volvo_Truck_Corporation.pdf.

Biodiesel data sheet

BATCH DETAILS

Batch #	Date:	Water %	Volume	Feedstock Source
	..... /..... /.....	%	L/Gal*	....

TRANSESTERIFICATION

In:	....	L/Gal* Methanol	In:	....	kg/lb* KOH
Out:	....	L/Gal* Glycerine			
In:	....	L/Gal* Methanol	In:	....	kg/lb* KOH
Out:	....	L/Gal* Glycerine			
In:	....	L/Gal* Methanol	In:	....	kg/lb* KOH
Out:	....	L/Gal* Glycerine			

AGITATION / MIXING

Start:	Time:	.... : .... AM/PM	Temperature:	.... °C/°F
Finish:	Time:	.... : .... AM/PM	Temperature:	.... °C/°F
Start:	Time:	.... : .... AM/PM	Temperature:	.... °C/°F
Finish:	Time:	.... : .... AM/PM	Temperature:	.... °C/°F

DEHYDRATION

Start:	Time:	.... : .... AM/PM	Date:	..... /..... /.....	
Finish:	Time:	.... : .... AM/PM	Date:	..... /..... /.....	
Start:	Water Content Before Dehydrator			....	%
Finish:	Water Content Before Dehydrator			....	%
Start:	% Atmospheric Humidity			....	%
Finish:	% Atmospheric Humidity			....	%
	Date In Finished Tank			..... /..... /.....	

WATER WASH

Wash No.	Wash Type	Date	Volume
Wash (1)	Mist/Light Heavy Agitation/With Acid*	..... /..... /.....	L/Gal*
Wash (2)	Mist/Light Heavy Agitation/With Acid*	..... /..... /.....	L/Gal*
Wash (3)	Mist/Light Heavy Agitation/With Acid*	..... /..... /.....	L/Gal*
Wash (4)	Mist/Light Heavy Agitation/With Acid*	..... /..... /.....	L/Gal*

*Delete As Applicable

OTHER OBSERVATIONS

Biodiesel and tax

Different countries have different procedures for taxing biodiesel and for charging excise on vegetable oils used as fuel. Ensure that you are familiar with the relevant legislation in your locale, to ensure that you are up-to-date. The information contained below, is subject to change: Check current rulings in your area; however, it serves as a guide to the procedures for declaring biodiesel production in various countries at the time of print.

United States

If you live in the United States, and you want to produce biodiesel, you must register with the IRS as a biodiesel producer using Form 637. Form 720 is then used to file a quarterly return on all Biodiesel produced. These forms can be downloaded from www.irs.gov.

United Kingdom

Until 2007 the law in the UK with regard to biodiesel, vegetable oil, and tax (duty) was confusing to say the least. One used to have to be registered as a fuel producer and return monthly reports to HM Revenue & Customs (HMRC) declaring how much biofuels one "set aside" (what constituted a "biofuel" and what constituted a "fuel substitute," which has a different rate of duty on it, was not at all clear) and enclose a check for the duty on it.

Now, so long as you don't produce more than 2500 liters in any rolling 12-month period you no longer need to do this. However, you are still required by law to keep a full and accurate record of fuels put into your vehicle and the mileage at the time.

More information can be found on the HMRC website:

www.hmrc.gov.uk

Australia

In Australia, the leaflet that gives you up-to-date information on paying excise for fuel is "Meeting Your Biodiesel Obligations NAT9885." However, there is also a scheme in Australia, the "cleaner fuel scheme," which offsets the duty payable, so that effectively, no excise is paid. This should operate until 2011, when it will be progressively phased out until 2015. Information can be found in leaflet, "The Cleaner Fuels Grants Scheme NAT9886," More information can be found at: www.atogov.au/excise or you can call 1-300-657-162.

Further reading

"How to make biodiesel."
Dan M. Carter, Jon Halle
Low-Impact Living Initiative
2005

"Biodiesel homebrew guide."
Maria "Girl Mark" Alovert
Local B100
localb100.com/book.html

"Biodiesel basics and beyond: a comprehensive guide to production and use for the home and farm."
William H. Kemp
Aztext Press
2006

"From the fryer to the fuel tank: the complete guide to using vegetable oil as an alternative fuel."
Joshua Tickell, Kaia Tickell, Kaia Roman
Tickell Energy Consultants
2000

"Biodiesel: growing a new energy economy," 2d ed.
Greg Pahl
Chelsea Green Publishing
2008

"Do it yourself guide to biodiesel: your alternative fuel solution for saving money, reducing oil dependency, and helping the planet."
Guy Purcella
Ulysses Press
2007

"Biodiesel power: the passion, the people, and the politics of the next renewable fuel."
Lyle Estill
New Society Publishers
2005

"Biodiesel America: how to achieve energy security, free America from Middle-East oil dependence and make money growing fuel."
Josh Tickell
Yorkshire Press
2006

Index

Page numbers for figures are given in **bold**

biodieselSMARTER

for biodiesel brewers by biodiesel brewers

Special magazine subscription discount offer to purchasers of this book

in aid of

CANCER RESEARCH UK

More than one in three of us will develop cancer at some point in our lives, many of us have lost someone to it, we all want cures to be found.

Cancer Research UK is the world's leading independent organization dedicated to cancer research.

They support research into all aspects of cancer through the work of more than 4,500 scientists, doctors, and nurses, benefiting cancer patients throughout the world.

Their vision of beating cancer is an enormous challenge. They are the leading independent funder of cancer research in Europe and they are almost entirely funded by the public – without the generosity and dedication of people like us they would not be able to continue to make progress in the fight against cancer.

Please consider making a donation.

http://www.cancerresearchuk.org

Cancer Research UK
P.O. Box 123
Lincoln's Inn Fields
London
WC2A 3PX
UK
Registered charity no. 1089464

in aid of

CANCER RESEARCH UK

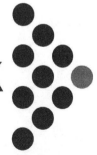